# WESTERN TIMES

BR (W)

CHURCHWARD SAINT CLASS SPECIAL
AUTUMN 2025

## Contents




The Transport Treasury

TIMES SERIES

Above: Colour photographs of 'Saint' Class locomotives, at any stage of their service careers, are a rare commodity. This particular example shows No. 2934 *Butleigh Court* stabled at its home shed of Swindon on 11 June 1950. The engine has recently received the neo-LNWR style of lined black livery that British Railways was then applying to its mixed traffic fleet. Whilst not in keeping with the traditional Swindon taste, it has to be argued that it sat well on the class. Although yet to receive an 82C shedplate, the SDN stencil above the cylinder confirms the allocation. The locomotive was to remain in traffic for a further two years. *R C Riley (RCRC 7264).*

Front Cover: The colourising of black and white images is a contentious topic amongst many, but due to the situation described above, is one of the only methods to appreciate how a 'Saint' would have looked in GWR regalia. An ex-works No. 2933 *Bibury Court* is seen at Swindon on 28 November 1945, an early recipient of green livery following the wartime austerity black. *David Williams.*

Rear Cover: No. 2938 *Corsham Court* is captured at Cardiff General on Thursday 15 May 1952. As can be seen by the 85C shedplate, the locomotive was allocated to Hereford at this time, following a spell at Gloucester since Nationalisation. It was to toil out its final months undertaking secondary passenger turns, being withdrawn in August 1952 having run 1,704,165 miles. *R C Riley (RCR 3571).*



ISBN 978-1-917776-15-8

First Published in 2025 by Transport Treasury Publishing Ltd.
16 Highworth Close, High Wycombe, HP13 7PJ.

Compiled and designed in the UK. Printed in Tarxien, Malta by Gutenberg Press Ltd.

# INTRODUCTION

The last decade of the 19th Century witnessed emergence of programmes sponsored by several operators to develop a new generation of express passenger locomotives. With expectations of heavier loads and faster speeds, reliance on single driver and four-coupled engines would be unsustainable. Long-term operational effectiveness would require of ten-wheeled, six-coupled locomotives and introduction of this genre was to prove chequered across the railway industry.

The GWR's success in tackling the enlargement challenge was the result of a fortunate chain of circumstances. Locomotive Superintendent William Dean found time, while brilliantly managing the unique complexity of gauge conversion and its aftermath, to explore fresh ideas and concepts. His enquiries were remarkable in their diversity ranging from personal experiments in metallurgy that preceded establishment of a fully-fledged laboratory at Swindon works to a varied range of unusual designs. History has deemed several as failures (e.g. tandem compounds, the Kruger family, and mysterious 4-2-4T No. 9) but the empirical knowledge gathered in the process was priceless in helping determine those ideas that might work and those best avoided.

The second stage in the chain was the harmonious and fruitful inter-generational partnership between Dean and his successor. Sometime before his retirement it was acknowledged who would be the next leader at Swindon and he was content to allow his deputy to pursue his own ideas in facing emerging demands. From about 1895 onward, there is doubt whether particular features of Great Western locomotive practice were attributable to Dean or to George Jackson Churchward, so intertwined was their technical relationship. The seamless nature of the transition on the cusp of seismic industry change was evident in Churchward's definition of the solution motive power needs, prepared in 1901 even before Dean's retirement.

The concluding stage was embedded in Churchward's character. Perhaps without matching his predecessor's intellectual stature, he was a superb leader which was crucial to the overall competence of the vast spectrum of manufacturing and operational activities for which was responsible. He could entertain alternative ideas even when they might conflict with his own preconceptions. Most beneficially he looked abroad for ideas, beyond the confines of a largely parochially-minded engineering community.

The speed and breadth of change in locomotive practice at Swindon before World War 1 was unprecedented anywhere before or since. Success embraced all of Churchward's creations, and although refinement would follow, the key essentials were designed, developed and proven in just three prototypes. In due course they were absorbed into a class of locomotives admired and revered by operator and enthusiast like. They were called the Saints.

--- o O o ---

Editor: Andrew Malthouse

Editorial Assistant: Jeremy Clements

To contact the editorial team please email: **WesternTimes@mail.com**

For sales and back issues of *Western Times* please go to: **www.ttpublishing.co.uk**

Above: Before the 'Great Awakening'. Joseph Armstrong designed 'Queen' Class 2-2-2 No. 1120 pictured at Bath *circa* 1890. The train is described as a down express although the composition suggests a secondary working. The load is modest compared with those that would typify train formations in the Edwardian era.

Below: This express train comprises ten or eleven vehicles of the late clerestory/ early elliptical period, led by a ten compartment non-corridor Diagram C23 58' All Third, followed by shorter clerestory Composite. While location and journey details are unknown, this duty must have been a tough challenge for a Dean 4-2-2. The identity of this 'Achilles' Class appears to be No. 3064 *Duke of Edinburgh*. The presence of post-clerestory coaches and the tender livery indicates a date toward the close of the locomotive's working life and the scorched smokebox door indicates recent hard work. The damaged front buffer beam is atypical of the condition in which these iconic locomotives were usually presented. The D0-type boiler shows how the wheel diameter prohibits use of a tapered boiler.

# LOCOMOTIVE ENLARGEMENT
## THE DEMANDS AND CHALLENGES

During the first decade of the 20th Century, there was a trend to replace the title of Locomotive Superintendent with Chief Mechanical Engineer, which better described the all-encompassing role of the chief executive officer in charge of a railway's mobile assets. In the case of G J Churchward, although the job content was effectively unchanged, the new title acknowledged the diverse activities for which he was ultimately responsible in such a large commercial enterprise.

His department employed over 37,000 personnel engaged in the manufacture and maintenance of locomotives, carriages and wagons at works, depot and shed level plus the efficient operation of that equipment. In geographic terms, the scale was enormous as the pre-Grouping Great Western was the largest UK railway company measured by route mileage. Annual expenditure under his control ran into millions (even in pre-Great War financial terms), deployed across capital investment and operational expenditure. Minimisation of cost while maximising operational efficiency was essential for the general health of the company, and for the benefit of its owners. Such obligations must have been in the forefront of the great man's mind when in January 1901 he released diagrams for six proposed 2-cylinder locomotive types as the core of his revolutionary plans for motive power development.

There were several novel features in these proposals that were particularly significant but three stand out. Firstly, Churchward's flagship designs were expected to achieve a combination of speed and haulage capacity that his peer group on other railways considered unattainable – until his creations' performances confounded the cynics. Secondly, this near fleet-wide generation of new locomotives would rely heavily upon design standardisation in a programme that only excluded the existing assembly of small but doughty six-wheeled tank and 0-6-0 tender locomotives. Thirdly, there was no 'dud' design among the new generation!

The six diagrams were initially designated alphabetically:

| | Wheel Arrangement | Driving Wheel Diameter | Boiler Standard Type | Classes |
|---|---|---|---|---|
| A | 2-8-0 | 4' 7.5" | No. 1 | 28xx |
| B | 4-6-0 | 5' 8" | No. 1 | Grange § |
| C | 4-6-0 | 6' 8.5" | No. 1 | Saint |
| D | 2-6-2T | 5' 8" | Nos. 2/4 | 31xx/ 3150 |
| E | 4-4-2T | 6' 8.5" | No. 2 | 'County' Tank |
| F | 4-4-0 | 6' 8.5" | No. 4 | 4-4-0 County |

§ Late arrival in 1936 but nonetheless a splendid performer.

All originally envisaged parallel boilers but this notion was soon rectified in favour of the part- and full-coned barrels of the standard range. Classes E and F could tenuously trace their ancestry back as advances upon earlier types (the 'Birdcage' 2-4-2Ts and the outside framed 4-4-0s respectively) but Classes A to D foreshadowed substantial advances in size and power.

Standardisation usually relates to locomotive groups that share familial resemblance which certainly typified GWR motive power. However, Churchward's principles had greater reach, embracing key components (e.g. ejectors, wheel diameters, boilers, superheaters) plus mundane items (e.g. nuts, bolts, rivets) of consistent dimensions regardless of the size of locomotive to which they were fitted. For logistical management the advantages were considerable and resilient. In BR days, personnel who had worked for others of the Big Four on employment rotation were surprised at the small range of spare parts needed at depots remote from Swindon. The benefits were profound, especially as the discipline extended throughout the CME's domain beyond merely the modern locomotive fleet.

Despite the new order's iconoclastic nature, Great Western design practice remained a constant, as evident in the standards set out in the adjacent table. In earlier years, driving wheel diameters had increased only by virtue of greater tyre thickness. Further, Churchward's focus on boiler development led to the obvious conclusion that a sound design could fit other classes of broadly similar size. Experimentation with 2-6-0 Class 26xx (the Aberdares) had explored the smaller/ larger and parallel/ coned debates with ultimate determination that the Standard No. 4, best met prevailing needs for intermediate-sized locomotives. This design, as iconic as the Standard No. 1, served with distinction both the Aberdare and its express passenger counterpart, the 4-4-0 City Class. It also found a home with eight-coupled freight and large prairie tank locomotives. Aberdares and Cities would soon be usurped from front line duties by arrival of larger locomotives but the Nos. 1 and 4 boilers (and variants) would continue in use until the mid-1960s.

Standardisation yielded a subtle benefit as the advantages of a full-coned boiler became apparent. Haulage demands upon a freight engine could be extreme but intermittent, interspersed by pauses to permit passage of faster traffic. In contrast, demands upon an express passenger locomotive were sustained continuously over longer periods. As the years passed, half-coned No. 1 boilers migrated from the early 4-6-0s to the Class 28xx, where they saw out their careers usefully.

## WESTERN TIMES SPECIAL

Although the 'partnership' with the 2-8-0s was a significant factor, the story of what became the Saint Class is the objective of this work. It is with these engines that Churchward's revolutionary ideas had the most profound impact upon professional railwayman and informed enthusiast alike. Much has been written of his considerable qualities but perhaps most appealing in this context was his avoidance of the parochial 'not invented here' syndrome that permeated the British mechanical engineering fraternity at the turn of the century. This open-minded attitude led to study of prevailing French and US engineering practice that proved particularly beneficial.

A combination of factors drove the need for more powerful passenger locomotives. Changing passenger expectations meant that long-distance journeys in non-corridor stock were unacceptable. This had profound impact upon tare weight per passenger seat with the need to provide for corridors, vestibules, gangway connections, and toilets. Covering the period from the 19th Century convertible era to the advent of steel-bodied bow-ended coaches, the table below illustrates trends in dimensions, tare weights and carrying capacity. These statistics are based on the number of All Third coaches required to accommodate a notional load of 300 seated passengers:-

| Diag. | Body Type | Built | No.of Seats | Body Length | Tare (Tons) | Tare per Seat | No. of Coaches | Aggregate Length | Total Tare (Tons) | Tare per Passenger |
|---|---|---|---|---|---|---|---|---|---|---|
| C3 | Clerestory Non-Corridor | 1890-93 | 80 | 46' 6" | 20.1 | 0.25 | 4 | 186' | 81 | 0.27 |
| C17 | Clerestory | 1898-03 | 64 | 54' 1" | 24.25 | 0.38 | 5 | 270' 5" | 121 | 0.4 |
| C24 | Dreadnought | 1905 | 72 | 69' | 32.45 | 0.45 | 5 | 345' | 162 | 0.54 |
| C27 | Concertina | 1906-7 | 80 | 70' | 32.85 | 0.41 | 4 | 280' | 131 | 0.44 |
| C29 | Toplight | 1909-13 | 80 | 70' | 33.55 | 0.42 | 4 | 280' | 134 | 0.45 |
| C32 | Toplight | 1914-5 | 64 | 56' 11" | 28.2 | 0.44 | 5 | 284' 7" | 141 | 0.47 |
| C54 | Steel body | 1925-29 | 64 | 58' 5" | 30.3 | 0.47 | 5 | 292' 1" | 152 | 0.51 |

**Above: The four members of the 'Armstrong' Class were introduced in 1894 as the first large-wheeled 4-4-0s on the narrow gauge. No. 7 *Armstrong* demonstrates the close resemblance of the type to the Dean 4-2-2s in original condition. Four-coupled 7' 1" driving wheels improved the adhesion element but performance was limited by the deign being under-boilered (or over-cylindered). This factor was a harbinger of the challenges in achieving correct proportions in the enlargement process. Bristol Temple Meads *circa* 1900.**

Above: William Dean's 'Aberdare' Class 2-6-0 No. 2610; this type could trace its lineage back to the Kruger 4-6-0/ 2-6-0 family and in the early years served as guinea pigs for Churchward's development work with the No. 4 boiler and its smaller No. 2 predecessor. Aping 4-6-0 No. 100, Aberdare No. 2662 carried a unique parallel No. 4 boiler before the tapered version became the standard for this 81-strong class, and for the large 2-6-2T, 2-6-0, 2-8-0T and 2-8-2T types that followed plus the 4-4-0 Counties.

Below: 'City' Class 4-4-0 No. 3441 *City of Winchester* hurries a Paddington to Oxford express through Southall on 24 April 1905. Fitted with the Standard No. 4 boiler, the Cities were the most potent and famous of the double-framed large-wheeled 4-4-0 family although their reign as the company' leading express type was quite short in face of the arrival of two and four-cylinder 4-6-0s. The relationship between the Cities and the Aberdare 2-6-0s was analogous to that of the Saints and Class 28xx.

A further factor, unique to the GWR was the need to establish on-train catering (where the tare/ passenger relationship was even more unfavourable) in conjunction with the costly termination of the notorious Swindon 'refreshment' rooms lease, effective from 12 November 1895. The unprecedented non-stop passage at speed of an ordinary passenger train through the station actually occurred on 1 October that year. Until then, there was no service on any route that included a non-stop run over 100 miles but general acceleration of timings soon commenced. These were examples of non-stop journeys:

| Train | Miles | Journey Time |
|---|---|---|
| The 'Cornishman' London-Bristol | 118.5 | 2 hrs 15 mins |
| South Wales Express London-Bath | 107 | 2 hrs |
| South Wales Express Newport-London | 143.5 | 2 hrs 58 mins |
| Down 'Cornishman' London-Exeter § | 194 | 3 hrs 45 mins |

§ Introduced July 1896, this unadvertised working was the longest non-stop journey in the country.

These improvements were possible with the new-fangled installation of water troughs between Pangbourne and Goring and between Keynsham and St. Anne's Park plus relevant tender modifications. The technology was readily understood as the London & North Western had started using troughs around 30 years earlier, a measure of how much ground had to be made up. Nevertheless, what MacDermot termed the 'Great Awakening' was under way.

### Wheel Arrangements for Express Services

The first sign of revolutionary change in motive power policy was expressed by Churchward in prototype 2-cylinder 4-6-0 No. 100 whose delivery in February 1902 personified the shock of the new, as reviewed in later chapters. At the time, the GWR's high speed services were in the hands of elegant Dean 4-2-2s of which all eighty were still in service, with various styles of boiler. Introduction of steam sanding had extended single drivers' operating lives. Nevertheless. they were increasingly out-dated against prevailing traffic demands and the type completely disappeared within thirteen years. Fast and smooth-running, considerable nostalgic sentiment attached to these engines, and even forward-looking Churchward explored how the lives of a select few might be extended with a more modern boiler. Such ideas foundered on the inescapable limitation of poor adhesion when starting with heavy loads and in hill-climbing. Further, the 7' 8" driving wheels prevented installation of a tapered boiler which foreshadowed the complexity of achieving the right 'balance' in design enlargement. The steam locomotive, for all the inherent inefficiency of an external combustion engine, is remarkable for the amount of haulage capacity generated from a volume constrained by rail and loading gauges.

**Above: 'Badminton' Class 4-4-0 No. 3297 *Earl Cawdor* was evidence of Churchward's open-mindedness. With the agreed policy favouring tapered boilers and Belpaire fireboxes under way, a member of the Swindon drawing office proposed that a large diameter parallel boiler with round-topped firebox could perform as well as and perhaps better than the classic Great Western form. The case must have been well argued as authority was granted for this locomotive to be rebuilt in July 1903 as illustrated here. For good measure, a spacious side-window cab was added which was definitely more comfortable that the Swindon standard of those days. No appreciable advantage was identified under test and the locomotive reverted to that of a normal Badminton with a Standard No. 4 boiler in October 1906.**

**Above: The most tangible example of Churchward's interest in overseas practice was his purchase of French de Glehn Compound No. 102 in October 1903 later named *La France,* plus two more larger examples in June 1905. Despite this design's excellent reputation and after extensive testing, Churchward opted for the 4-6-0 wheel arrangement with simpler transmission and single expansion, as vindicated by Saint No. 171 *Albion* in temporary 4-4-2 form. The key feature retained from the Frenchmen was four-cylinder divided drive. Nos. 103/ 4 were fitted with side window cabs which were eventually adopted in Collett's time.**

An advantage of the single driver was the minimal number of mechanical parts employed in power transmission. The next mechanical stage in enlargement i.e. four driving wheels added the necessity of coupling rods. On other railways during workshops visits, measurement of driving wheel diameters for early six-coupled locomotives with centre driving axles revealed that typically, tyre wear was most acute with the trailing driving wheels. Data appears unavailable for GWR locomotives and this might have been an insignificant factor due to Swindon's superior engineering standards but uneven tyre wear was indicative of the adverse impact of internal frictional resistance. Attempts to mitigate this element helped explain why initial emphasis was on four-coupled designs whereas so far as the GWR was concerned, the 4-4-0 fulfilled a valuable interim role pending refinement of larger locomotives.

Focus elsewhere was on enlargement using 4-4-0 or 4-4-2 types with the option of compound steam, before superheating achieved pre-eminence. The re-cycling of expended steam was theoretically attractive in pursuit of improved transmission and Churchward conducted his own investigations through acquisition of de Glehn Compound No. 102 *La France* in October 1903. His subsequent judgement that its transmission was unnecessarily complex was seminal (although he did revive the investigation with two more de Glehn compounds later). Instead, he had already favoured the much simpler configuration of two cylinders with traditional Stephenson's Link Motion. This choice was profound as it amply complemented the superb No. 1 boiler that had been the prime focus of his initial developmental efforts. This feature was first apparent in his experimental 4-6-0 No. 100, and was refined with No. 98, the first Saint proper.

The Churchward revolution thus commenced with Nos. 100 and 98. The pace of its implementation and the breadth of its success was remarkable, as was also the extended interval before other railway operators acknowledged the advantages that Churchward's work had bestowed upon the GWR. In this context, milestones over 62 years apart are worth considering. The year 1903 saw the arrival in service of No. 98 and a search for another railway's 4-6-0 that matched its performance will confirm how far ahead of the other operators was the GWR in the Edwardian era. The date 31 December 1965 marked the formal end Great Western steam. This was a dismal episode that Swindon devotees of a certain age would prefer to forget, but No. 4920 *Dumbleton Hall* was one of the last withdrawals that month. No. 4920 was the final survivor of the first production batch of Halls and in spite of its modern cab and smaller driving wheels was by way of No. 2925/ 4900 a Saint re-incarnate.

# INITIAL DESIGN AND CONSTRUCTION

By 1900, there was general recognition among enthusiasts and the more discerning element of the travelling public of the traditional style and appearance of the Great Western express passenger locomotive. This glittering machine might lack admiration from enthusiasts whose interests lay elsewhere but few would deny that the fleet looked fundamentally different from those of other railways. The oft-quoted reference by the lady to the 'engines that look like knights in armour' aptly summarised the distinction.

Later generations, for whom the 2-cylinder 4-6-0 was a constant and much admired feature of the railway, might thus be surprised at the adverse reaction induced by introduction of No. 100. Mr Churchward's hand had been evident with earlier locomotives but his experimental work thus far had focussed on boiler design. Setting aside the aberration of the ungainly 'Kruger' and his family, insertion of the coned Belpaire boiler had been unobtrusive even against the stylish curved running plates of the Badmintons, Dukes and early Bulldogs. By comparison, No. 100 in its gaunt, angular, utilitarian appearance was an affront to late-Victorian and Edwardian sensibilities.

This functionalism had purpose as the company's previous 4-6-0s were transitional. No. 36 (the 'Crocodile') was large by the standards of the time, imposing in the Dean design tradition but redolent of 19th Century practice. No. 2601 ('Kruger') exemplified how broadly Churchward had cast his net in the search for enhanced boiler performance. Visually this creation was dominated by an unsightly boiler, married to a traditional layout below the running plate.

The earliest examples of the locomotives that would ultimately form the template for the Saints deserve individual treatment.

### THE EXPERIMENTAL TRIO

**No. 100 (later No. 2900)**
**Built February 1902 under Lot No. 132**

The pioneer two (outside) cylinder 4-6-0 set out a fresh design philosophy that was quite unlike anything seen hitherto on the GWR. It is easy to appreciate the shock of the new in this progenitor of a fresh express motive power generation. The running plate was set high to accommodate the 6' 8.5" driving wheels and its form was in distinct contrast with the flowing lines of the Dean era. The front fall was achieved by a pair of reverse right angles to enable setting of the buffer beam at the requisite height above rail level. The difference between the main running plate and the forward section above the bogie was comparatively shallow and in compensation the buffer beam was deeper. The main running plate continued back to the rear of the footplate on the same level and three substantial footsteps gave access to the cab. The height above rail level was accentuated by the absence of protective outside frames and the fully exposed nature of all locomotive wheels. The ensemble lacked traditional symmetry in the height difference between locomotive and tender running plates.

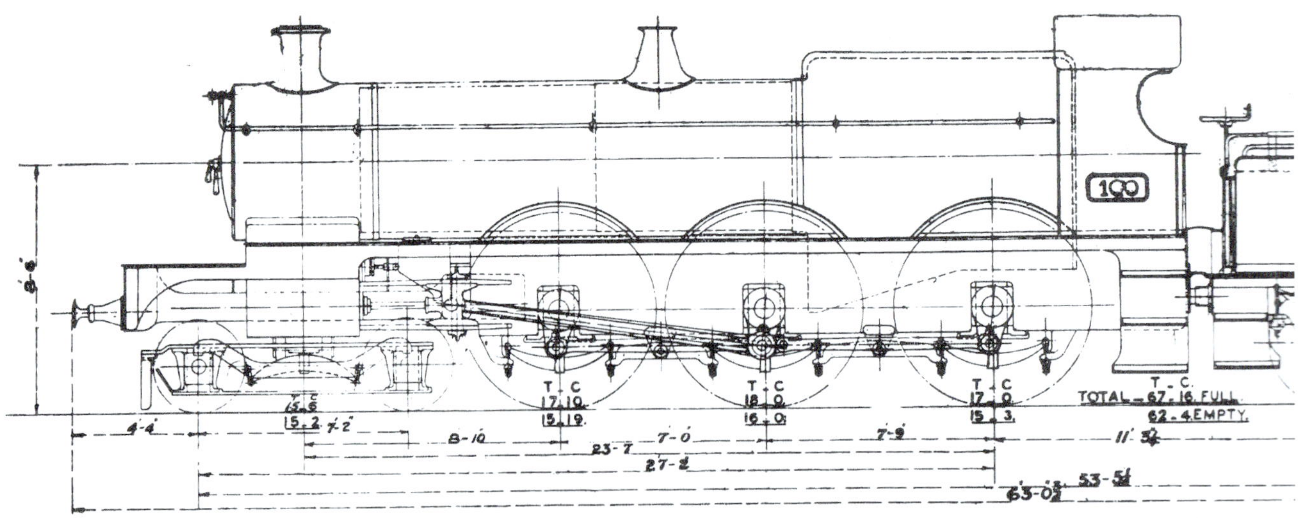


**Above: Drawing of No. 100 in original condition with the short, smokebox experimental parallel boiler and steel Belpaire firebox. The unique cylinder form remained in place throughout the locomotive's career. Having reached the cylinders' re-boring limit, this was the reason for withdrawal in 1932 although by then other early examples were in process of being taken out of service.**

**Churchward 4-6-0 No. 100 in as-built condition and prior to its first naming as *Dean* in June 1902. The location is Westbourne Park shed. *Stephenson Locomotive Society.***

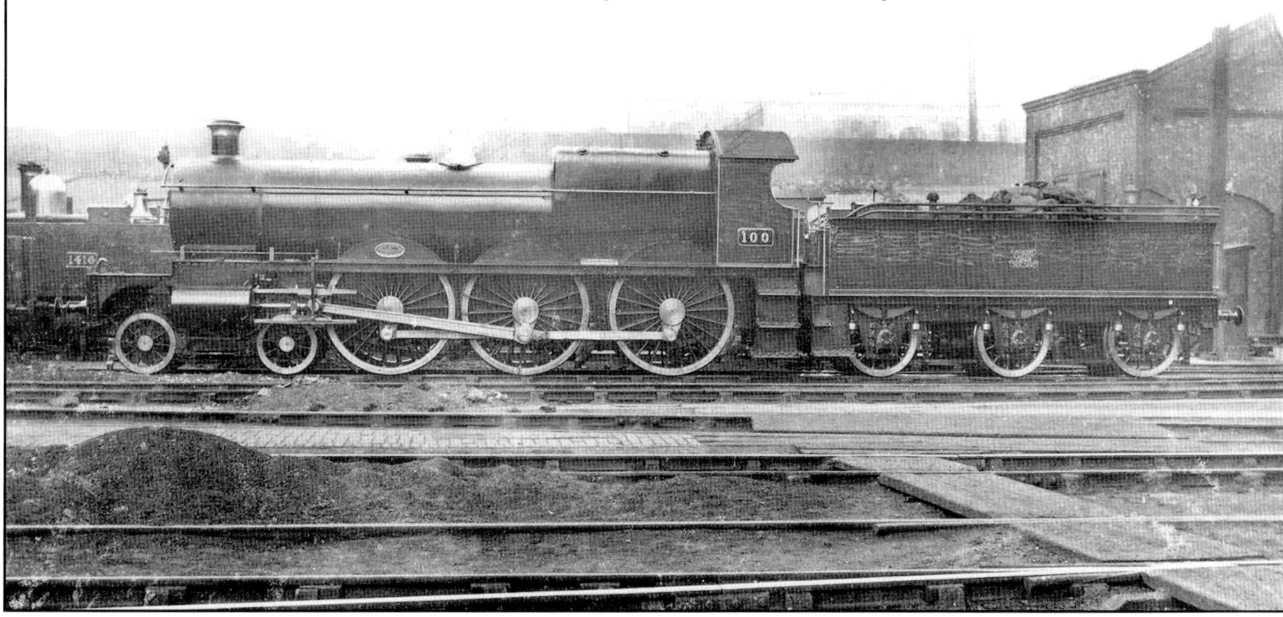

The Spartan cab certainly accorded with earlier designs while the angular bulk of the Belpaire firebox was exaggerated by the parallel boiler, the first example of the famous Swindon No. 1 type. The whole appearance portrayed stark functionalism which was distinctly at odds with fashionable ideas on the style that should be employed in an important GWR type. Retention of standard livery [mid-chrome green, Indian red below the running plate plus copper and brass) and attachment to a normal tender did little to assuage the objectors.

The parallel boiler was experimental and unique to this locomotive, reflecting that Churchward's investigation into maximisation of steaming capacity which had commenced under Dean was still in its evolutionary phase. Parallel versions of the Standard No. 2 boiler were under construction between 1899 and 1902 for use with 4-4-0s (mainly Bulldogs) and Aberdare 2-6-0s Nos. 2600/ 21-60 (new construction). Churchward parallel boilers were transitional as they lacked the extra steam-raising capacity provided by the coned shape. The vessel on No. 100 was replaced in June 1903 with a half-cone boiler which in turn gave way to the fully coned version in February 1912 which type was carried until withdrawal. A dimensional summary of the various forms of the Standard No. 1 boiler used with the Saints appears at the end of this chapter.

This locomotive's other unique features concerned frames, cylinders and valve gear as a further manifestation of Churchward's evolutionary work. The double ported 6½ inch diameter piston valves were mounted on the inside of the frames and the valves were actuated independently. The valve diameters were later increased to 10 inches.

Withdrawals of Saints commenced in late 1931 and the unique cylinder layout of No. 100 (by then No. 2900) meant that this pioneer would be an early candidate for removal from service. This occurred in June 1932 when the unique cylinders had reached their re-boring limit.

When first named, the single word *Dean* appeared on a long plate that was adjacent to the full span of the leading splasher with an oval works plate below on the splasher face. On renaming as *William Dean*, the conventional style of plate was mounted on the centre splasher and with the usual spacer. This gesture was highly appropriate as Churchward had enjoyed a long, harmonious and fruitful relationship with his predecessor whose experimental work in earlier years provided invaluable empirical evidence to aid locomotive enlargement at commencement of the new century.

**No. 98 (later No. 2998)**
**Built March 1903 under Lot No. 138**

Churchward's continuing focus on steaming efficiency saw installation of the first Standard No. 1 boiler with this locomotive. It was of the half or short cone variety combined with a firebox that tapered inwards toward the cab. With the full-cone boiler that was later acquired, the resultant profile set the form for the multitude of 4-6-0s and 2-8-0s that would follow. Like that on No. 100, the working pressure was 200 lb/ sq in.

No. 98's other major fresh feature lay in chassis design. Each cylinder was cast as a single unit that included one steam chest and half the smokebox saddle. The layout was symmetrical so that the two cylinder units were bolted back-to-back. The resulting structure was very robust and significantly reduced points at which steam leakage might occur. The symmetry meant that over time with uneven wear in the bores, compensation was easily affected by reversing the cylinder units. To accommodate this structure, the conventional mainframes terminated immediately behind the cylinders and bar frame extensions were bolted to the plate frames. The cylinder assembly was bolted to the bars.

No. 98 was built concurrently with pioneer 2-8-0 No. 97 and an issue arose concerning tyre wear that over time might lead to breach of the loading gauge. Rather than adopt inclined cylinders which were anathema to Churchward, in compensation the cylinder centre line was off-set by 2½" above that of the driving axle. This feature, later deemed unnecessary, was confined to the earlier engines but re-appeared with the last Saint to be built.

The quest for steam efficiency embraced design of the piston valves whose diameter was 10", large by contemporary standards, with the long travel of 5 7/8" that was later increased to the standard of 6¼". The use of Stephenson's Link Motion gave the 2-cylinder 4-6-0 family extra 'punch' on starting and in hill-climbing which was why later smaller-wheeled classes were later so favoured in areas like Cornwall.

This, the first true Saint graduated to a fully tapered boiler in November 1908 but reverted to the short-cone version between January 1913 and January 1917. The locomotive then remained unchanged until withdrawal in June 1933.

**Below: Churchward 4-6-0 No. 98 which was the first Saint proper, entered service in March 1903. It was appropriately named *Vanguard* in March 1907, but became *Ernest Cunard* the following December. *Stephenson Locomotive Society.***

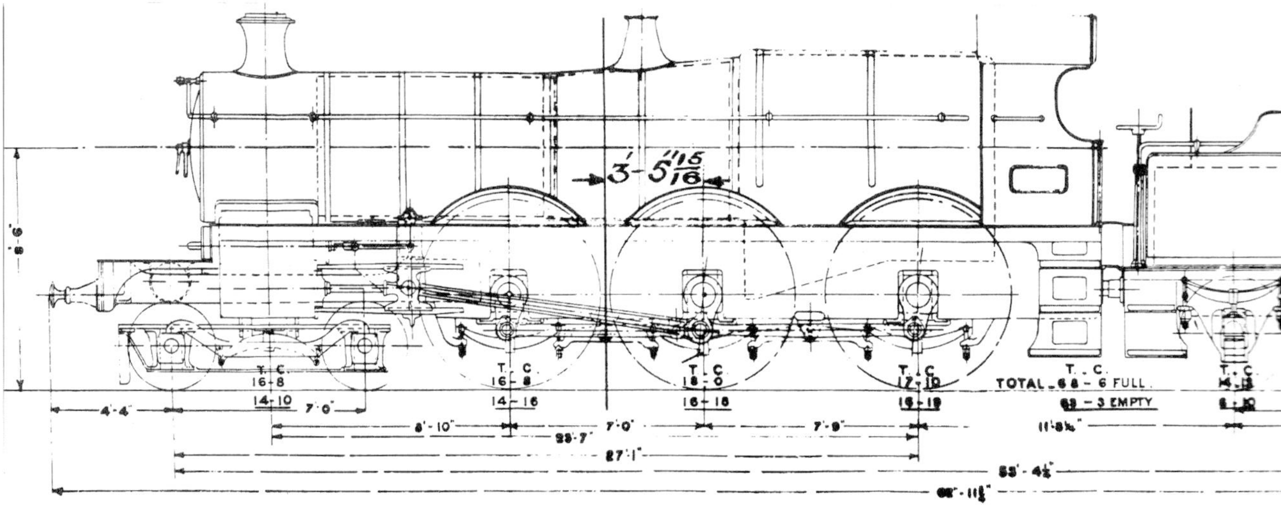


**Above: No. 98 which was the GWR's second outside-cylinder 4-6-0 differed on entering service with a short-cone saturated tapered boiler. The front end combination of bar fames bolted to plate frames immediately aft of the cylinders, set the pattern for the remainder of the Saints, plus the Halls, Granges and Manors of the Collett era. The shallow smokebox used with No. 100 was also fitted to No. 98.**

**No. 171 (later No. 2971)**
**Built December 1903 under Lot No. 145**

This locomotive was similar to No. 98 when built except that its short-cone boiler was pressed to 225 lb/ sq in which became a standard. There were also small amendments to the firebox dimensions which resulted in minor changes in weights. More profound were changes to the chassis after less than eight months in service.

At time of construction, all eighty of Dean's magnificent 'Achilles' Class 4-2-2s were still working express passenger services over less hilly routes. Their speedy, smooth-running abilities remained to the fore in the minds of operators and these qualities were in part attributed to the reduced level of internal mechanical friction in a single driver. In this context, the freer running characteristics of four-coupled engines were sufficient to support the contention that Atlantics might allow for a viable step-up in size for the longer term.

From today's perspective the idea might seem unjustified but there was sufficient evidence to indicate that 4-4-2s deserved further evaluation. The performance of compound locomotives with this wheel arrangement had persuaded Churchward to check for himself through purchase of No 102 *La France* which had entered service three months before No. 171. Closer to home, there were parallels with J G Robinson on the Great Central whose investigations also took him into Atlantic/ 4-6-0 and simple/ compound comparisons. His general conclusions were contrary to those of Churchward as his large 4-4-0s (the 'Directors') and 4-4-2s (the 'Jersey Lilies') proved more effective than his rather heterogeneous 4-6-0 family. Broadly speaking, GCR passenger loadings were lower and the superbly graded London Extension presented less of a challenge for four-coupled power.

In any event, comparative evaluation against *La France* included conversion of *Albion* to a 4-4-2 in which condition it ran from October 1904 until July 1907. The trailing coupled axle was replaced with a carrying axle and 4' 1½" wheels. Outside axle boxes were mounted on a short extra frame that was bolted to the corresponding section of the main frame. The carrying axle was set 6" closer to the cab which increased the locomotive's overall wheelbase by that amount. The modification added rather more than two tons to the overall weight; the adhesive weight reduced from 51.9 tons as a 4-6-0 to 39 tons as an Atlantic. For comparison, the Frenchmen's adhesive weights were No. 102 [33.35 tons] and Nos. 103/ 4 [37.2 tons]. On those grounds, evaluation against the latter pair which entered service in June 1905 was probably more useful.

**Above: A contemporary postcard produced by the Great Western Railway to promote the release from Swindon of its latest Churchward 4-6-0. What purports to be No. 171 *Albion* is captured wearing the obligatory 'works grey' paint scheme for her construction portrait. There is doubt over the actual identity of the locomotive utilised, due to obvious doctoring of the name/ number and the position of the Works Plate on the central splasher, as No. 171 ran in service with it sited below the smokebox door.** ***GWR Official.***

## THE PRODUCTION BATCHES

### First Production Series: Lot Nos. 154 and 158 (all introduced 1905)

Delivered during the evaluation period vis-à-vis the French compounds, Churchward was evidently undecided concerning the wheel arrangement issue as Nos. 172/ 179-90 were built as Atlantics and Nos. 173-8 as 4-6-0s. Reversion of the first Atlantic (No. 171) to six-coupled status in July 1907 probably marked the decision point. Train loads were then increasing with introduction of 70' coaching stock (Dreadnoughts from 1904 and Concertinas from 1906) with consequential rise in tare per seated passenger.

All nineteen examples of the first production series were built with short cone boilers and only Nos. 179/ 80/ 8 retained this type through to conversion to 4-6-0s. Atlantic No. 187 acquired a convex front fall plate, apparently solely for aesthetic reasons to soften the front end profile in belated response to public objections over the stark, utilitarian appearance of the early locomotives. It is unclear whether this was installed during or post construction. The safety element of this appendage was similar to that which arose in the 1930s with the convex 'streamline' front running plate enclosure with Nos. 5005 and 6014.

A more comprehensive approach was adopted during rebuilding from Atlantic condition which was undertaken between April 1912 and January 1913. The abrupt angular running plate profile was replaced with the more pleasing front and rear curved fall plates that had been adopted with the third production series (Nos. 2911-30 built in August/September 1907 under Lot No. 170). Those engines which had always been 4-6-0s (Nos. 173-8) remained unchanged in that respect.

With effect from late December 1912, a renumbering scheme for all the early locomotives was instituted. Nos. 100/ 171-90 had 2800 added to their original numbers while No. 98's identity was increased by 2900. Thus all 2-cylinder 4-6-0s then in service were combined into the 29xx series; it seems that 'Saint' as the collective class title was then coined. The intricacies of the individual, series and class identities are discussed in the next chapter.

**Above: No. 181 *Ivanhoe* is pictured at Swindon around February 1907, following overhaul and naming. Built as a 4-4-2 in June 1905 and delivered to Newton Abbot, it ran as an Atlantic until conversion to a 4-6-0 in July 1912. Withdrawal came in March 1951. *GWR Official.***

### Second Production Series: Lot No. 164 (all introduced in 1906)

The ten locomotives of this series (Nos. 2901-10) and all subsequent new construction in the 29xx number block were 4-6-0s throughout their careers. Consistent with earlier locomotives, they carried the usual mixture of boiler types. Nos. 2901/ 4-6 appeared with the short-cone type while Nos. 2902/ 3/ 7-10 carried the long-cone version ex-works. There followed the usual pattern of changes and reversions except for Nos. 2907/ 8 which were long-cone engines throughout their careers. Withdrawals commenced in late 1931 although five (Nos. 2902/ 3/ 5/ 6/ 9) survived into BR(W) ownership.

### Third Production Series: Lot No. 170 (all introduced in 1907)

All introduced with long-cone boilers and named after Saints, Nos. 2911-30 comprised the largest individual Lot and construction was completed in August/ September. As mentioned previously, this series was the first to appear with the more pleasing curved fall plates front and rear. However, further variety was evident through boiler rotation as all but Nos. 2922-4/ 8/ 30 later acquired the earlier short-cone type at intervals later during their careers. The first withdrawal occurred in 1924 when No. 2925 was converted into No. 4900 as the prototype for the 'Hall' Class while the first conventional service withdrawal took place in early 1932. Eleven (Nos. 2912/ 3/ 5/ 6/ 20/ 4/ 6-30) were still in service at Nationalisation on 31 December 1947.

**Left: No. 2929 *Saint Stephen* is captured at the head of an eclectic express formation *circa* 1912. The locomotive and coaches are in typically immaculate condition, with all brass and steel fittings clearly having had recent attention from the shed cleaners.**

**Introduced as standard on this batch of locomotives were curved valances at the ends of the running plate and an inclined screw reversing lever. Also of note is the larger diameter chimney with capuchon and the retention of the oval works plate on the centre splasher.**

**The tender is Churchward 3,500 gallon No. 1753, built to Lot. A75 in October 1907.**

### Fourth Production Series: Lot Nos. 185, 189 and 192 (introduced between 1911-13)

After a four-year hiatus, construction resumed at a less frenetic pace. Diminishing availability of short-cone boilers was apparent in that only five (Nos. 2931/ 5/ 9/ 42/ 3) carried this type later in their careers. Apart from No. 2901, these Lots comprised the only class members to be superheated from new. Except for No. 2947, they later acquired outside steam pipes and all survived to become BR(W) property.

| Lot No. | Nos. | Year | Introduced |
|---|---|---|---|
| 185 | 2931-40 | 1911 | Oct to Dec |
| 189 | 2941-50 | 1912 | May/ Jun |
| 192 | 2951-55 § | 1913 | Mar/ Apr |

§ This lot originally consisted of ten locomotives but the construction of Nos. 2956-60 was cancelled.

**Above: The final production 'Saint' was No. 2955 *Tortworth Court*, which was released to traffic shortly after this 'Works Photograph' was taken in April 1913. Its first allocation was to Newton Abbot and it was to have a service life of 37 years.** ***GWR Official.***

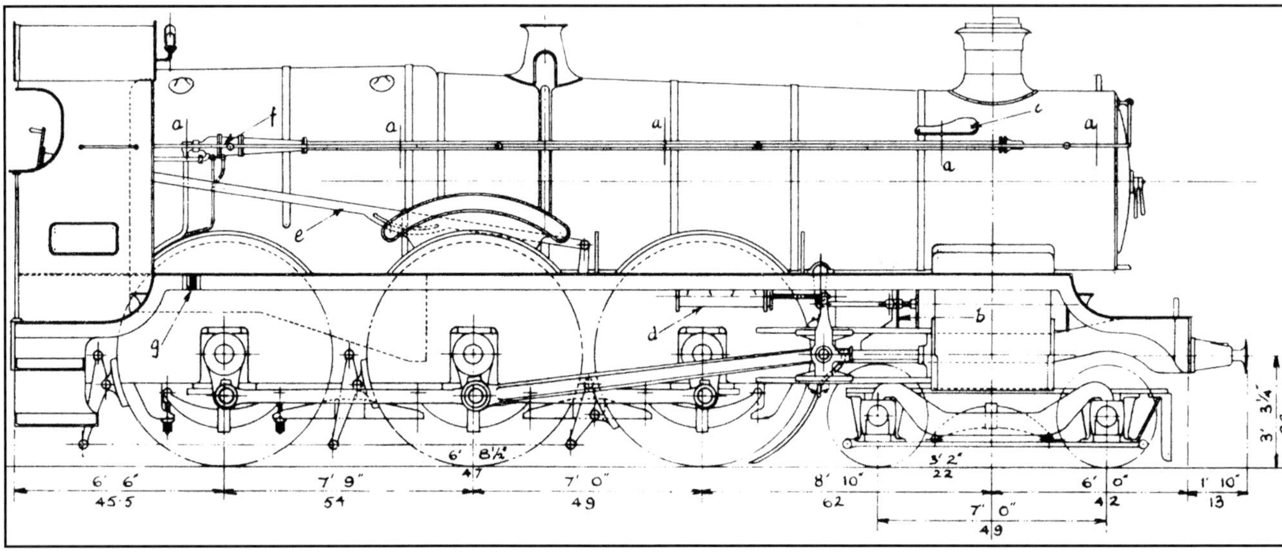


Above: Driver side drawing of a late condition (fourth production series) 'Saint'. Of note is the larger diameter chimney with capuchon, superheated boiler, top feed fitted safety valve bonnet and inclined screw reverser lever. The driving wheel diameter was 6' 8½", with a total wheelbase of 27' 3". Weight of the locomotive was listed at 69.4 tonnes. Outside steam pipes (not shown here) were first fitted to No. 2903 *Lady of Lyons* in November 1930.

## Boiler and Firebox Dimensions

With small variations in fractions of inches, leading part/ full cone boiler dimensions remained constant.

| Firebox | | Boiler | |
|---|---|---|---|
| - length | 8' 2" > 8' 3" | - length | 14' 10" |
| - width | 4' 9" | - width | 4' 11" > 5' 6" |
| - height | 6' 6" | | |

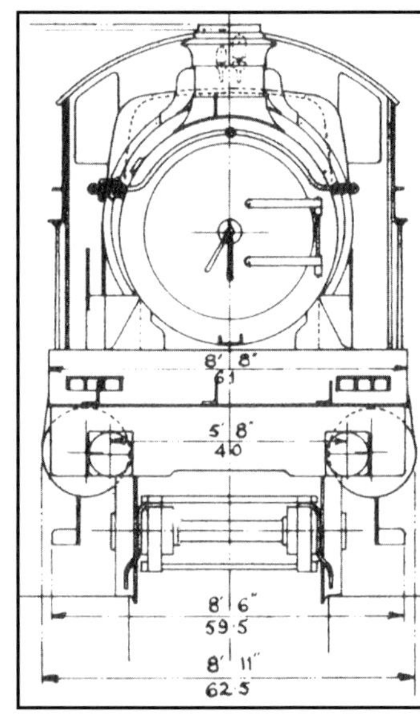


Right: This front elevation drawing highlights the neat proportions of the 'Saints' from pretty much all angles.

Principal dimensions were: 63' 0¼" (length), 8' 11" (width), and 13' 3½" (height).

Cylinder diameter was 18½" with a stroke of 30".

## Tube & Flue Dimensions; Heating Surfaces

| Boiler Type | Part Cone Saturated | Full Cone Saturated | Full Cone Superheated | Full Cone Superheated |
|---|---|---|---|---|
| **Number of Tubes** (2" diameter) | 250 | 247/ 250 | 166/ 176 | - |
| **Number of Tubes** (1.75" diameter) | - | - | - | 145 |
| **Superheat Flues** (5.125") | n/a | n/a | 14 | 21 |
| **Heating Surfaces** (sq/ft) | | | | |
| - tubes 2" | 1988 | 1966 > 1988 | 1599 > 1686 | - |
| - tubes 1.75" | - | - | - | 1583 |
| - firebox | 154/ 155 | 154/ 155 | 150/ 155 | 150/ 155 |
| - superheater | n/a | n/a | 270 > 330 | 315 |
| **Grate** (sq/ft) | 27.1 > 27.2 | 27.1 | 27.1 | 27.1 |
| **Quantity** | 54 **[a]** | 78 **[b]** | 792 | 136 |
| **Build Period** | Mar-03 > Dec-05 | Apr-06 > Jun-09 | May-09 > 1944 | 1944 > 1960 |

Notes: **[a]** Slight variations with one boiler built June 1903.

**[b]** Four built in 1903 with 301 x 1.75" tubes.

# INDIVIDUAL IDENTITIES
## LOCOMOTIVES AND SERIES

Prior to World War 1, the complexity of type and individual locomotive identification warrants the following staged explanation.

### Works Numbers

These were allocated to those which ultimately became locomotives in the 29xx series:

| Running Nos. | Works Nos. | Running Nos. | Works Nos. |
|---|---|---|---|
| 100 | 1928 | 2901-10 | 2199-208 |
| 98 | 1990 | 2911-30 § | 2259-78 |
| 171 | 2024 | 2931-40 | 2426-35 |
| 172/ 9-90 | 2106/ 13/ 14/ 28-37 | 2941-50 | 2476-85 |
| 173-8 | 2107-12 | 2951-55 | 2506-10 |

§ No. 2925 retained Works No. 2273 on conversion to Hall Class.

No. 100 entered service with the traditional oval works plates uniquely mounted on the leading driving wheel splasher faces. This arrangement was necessary because the vertical front fall plate on this non-standard front end design was too shallow to accept the works plate. Locomotives Nos. 98/ 171-90/ 2901-10 were built with the angular vertical fall plate on which a single works plate was installed centrally below the smokebox.

Thirteen locomotives (Nos. 172/ 9-90) were built new as 4-4-2s and were converted to 4-6-0s with curved framing front and rear between April 1912 and January 1913. With replacement of the vertical front fall, substitute works plates were affixed to the central driving wheel splasher face. As all class members up to No. 2910 were equipped with pole reverse, the horizontal lever arm set on the right hand side largely hid the works plate from view. No. 171 was built as a 4-6-0 and two photographs from its period before temporary conversion to 4-4-2 (October 1904 to July 1907) depict the locomotive with a works plate below the smokebox. An official view in photographic grey (see Page 13) with no plate in that position but with one on the centre splasher face is thought unreliable as the identity seems to have been 'doctored'.

A photograph exists of No. 4038 *Queen Berengaria* (built January 1911 with splasher works plate) alongside No. 2934 *Butleigh Court* (built November 1911 with no plate) which indicates that the practice of so adorning new locomotives ceased during that year. An archive of around four-hundred 'Saint' photographs (mainly undated) has been scrutinised, from which the following definite siting of works plates can be confirmed:-

| | 4-4-2 | 4-6-0 |
|---|---|---|
| **Front Fall Plate** | 172/ 4/ 6/ 9/ 180-7/ 9/ 90 | 171/ 2/ 4/ 5/ 7/ 8/ 86/ 2901-4/ 6-9 |
| **Centre Splasher** | n/a | 98/ 2914/ 5/ 7/ 20-3/ 5/ 9 |

It is surmised that those locomotives already with splasher plates retained them after 1911, and probably lost them together with the removal of splasher and cab beading as part of economies during the Great War. No examples have been traced of plates remaining from 1919 onward.

**Centre splasher detail on a 5" gauge model of No. 2921 *Saint Dunstan*, showing the positioning of the works plate upon the face.**

### Styles of Number and Name Plates

Number plates in standard 2, 3 and 4 digit styles were used. The backplate measured 11¾" (H) and between 15¼" and 25¼" (W). Digits were 5¾" (H) with a thickness of ½".

**Above: Standard GWR 4-digit numberplate from *Butleigh Court*.**

The naming policy for the experimental trio and first & second production series of locomotives (Nos. 98, 100, 171-90, 2901-10) was haphazard. The timeline of naming locomotives is detailed as follows:

| No. | Built | Named | No. | Built | Named |
|---|---|---|---|---|---|
| 100 | Feb-02 | Jun-02 | 185 | Jul-05 | Feb-06 |
| 98 | Mar-03 | Mar-07 | 186 | Jul-05 | Sep-06 |
| 171 | Dec-03 | Feb-04 | 187 | Aug-05 | Nov-05 |
| 172 | Feb-05 | As built | 188 | Aug-05 | 1907 |
| 173 | Mar-05 | As built | 189 | Sep-05 | Sep-06 |
| 174 | Mar-05 | As built | 190 | Sep-05 | 1906 |
| 175 | Mar-05 | 1907 | 2901 | May-06 | Oct-06 |
| 176 | Apr-05 | Apr-07 | 2902 | May-06 | Apr-07 |
| 177 | Apr-05 | Apr-07 | 2903 | May-06 | Apr-07 |
| 178 | Apr-05 | As built | 2904 | May-06 | Apr-07 |
| 179 | Apr-05 | As built | 2905 | May-06 | Apr-07 |
| 180 | May-05 | Mar-07 | 2906 | May-06 | May-07 |
| 181 | Jun-05 | 1907 | 2907 | May-06 | Apr-07 |
| 182 | Jun-05 | Sep-06 | 2908 | May-06 | May-07 |
| 183 | Jul-05 | Sep-06 | 2909 | May-06 | May-07 |
| 184 | Jul-05 | As built | 2910 | May-06 | May-07 |

No. 100 was introduced at a time when various nameplate styles were being applied experimentally to other classes. Having initially ran with number only, its first name (*Dean*) was applied in June 1902 to a plate of plain rim style affixed to the leading splasher face and which extended through its full arc. The following November, the name was changed to *William Dean* but the style of plate on the leading splasher face remained unchanged. In 1904, new plates were fitted in what became the standard style but still on the leading splashers and with narrow risers. The final change took place in February 1906 when the nameplates were moved to the centre splashers with normally proportioned risers.

The second 2-cylinder 4-6-0 to acquire a name was No. 171 in February 1904. In this case standard form nameplates (*Albion*) were fitted but snug to the curve of the centre splashers. From March to July 1907, this locomotive was renamed *The Pirate*, and is believed to have carried snug-fitting standard plates. On reversion to *Albion*, the original nameplates were restored and remained unique to this locomotive for the remainder of its career. The first three Atlantics Nos. 172-4 were built in February/ March 1905 and the third of this group carried the snug-fitting nameplate *Magnet* until its renaming in March 1907. No other nameplates of this type have been traced.

Nos. 2911-55 all carried nameplates in the standard style from date of construction.

**Above: No. 179 *Magnet* is pictured at Bristol Temple Meads shortly prior to its renaming as *Quentin Dunward* in March 1907. This locomotive was from the batch of thirteen locomotives constructed as 4-4-2 Atlantics, and demonstrates the as-delivered short cone boiler, small smokebox and narrow chimney. Note also how the nameplate sits flush to the splasher without a riser plate.** ***F W Brewer.***

**Names, Name Changes and Related Dates**

| No. | As Built | Later Changes |
|---|---|---|
| 100 | Unnamed | *Dean* (Jun 02)<br>*William Dean* (Nov 02) |
| 98 | Unnamed | *Persimmon* (1906)<br>*Vanguard* (Mar 07)<br>*Ernest Cunard* (Dec 07) |
| 171 | Unnamed | *Albion* (Dec 04)<br>*The Pirate* (Mar 07)<br>*Albion* (Jul 07) |
| 172 | *Quicksilver* | *The Abbot* (Mar 07) |
| 173 | *Robins Bolitho* | n/a |
| 174 | *Barrymore* | *Lord Barrymore* (May 05) |
| 175 | Unnamed | *Viscount Churchill* (Mar 07)<br>*Sir Ernest Palmer* (Feb 24)<br>*Lord Palmer* (Oct 33) |
| 176 | Unnamed | *Winterstoke* (Apr 07) |
| 177 | Unnamed | *Robertson* (Apr 07) |
| 178 | *Kirkland* | *Charles J. Hambro* (May 35) |
| 179 | *Magnet* | *Quentin Durward* (Mar 07) |
| 180 | Unnamed | *Cœur de Lion* (Mar 07) |
| 181 | Unnamed | *Ivanhoe* (Mar 07) |
| 182 | Unnamed | *Lalla Rookh* (Sep 06) |
| 183 | Unnamed | *Red Gauntlet* (Sep 06)<br>*Redgauntlet* (Jun 15) |
| 184 | *Churchill* | *Viscount Churchill* (Sep 06)<br>*Guy Mannering* (Apr 07) |
| 185 | *Peverill of the Peak* | *Winterstoke* (Feb 06)<br>*Peverill of the Peak* (Apr 07) |
| 186 | Unnamed | *Robin Hood* (Sep 06) |
| 187 | Unnamed | *Robertson* (Nov 05)<br>*Bride of Lammermoor* (Apr 07) |
| 188 | Unnamed | *Rob Roy* (Apr 07) |
| 189 | Unnamed | *Talisman* (Sep 06) |
| 190 | Unnamed | *Waverley* (Sep 06) |
| 2901 | Unnamed | *Lady Superior* (Oct 06) |
| 2902 | Unnamed | *Lady of the Lake* (Apr 07) |
| 2903 | Unnamed | *Lady of Lyons* (Apr 07) |
| 2904 | Unnamed | *Lady Godiva* (Apr 07) |
| 2905 | Unnamed | *Lady Macbeth* (Apr 07) |
| 2906 | Unnamed | *Lady of Lynn* (May 07) |
| 2907 | Unnamed | *Lady Disdain* (Apr 07) |
| 2908 | Unnamed | *Lady of Quality* (May 07) |
| 2909 | Unnamed | *Lady of Provence* (May 07) |
| 2910 | Unnamed | *Lady of Shalott* (May 07) |
| 2911 | *Saint Agatha* | n/a |
| 2912 | *Saint Ambrose* | n/a |
| 2913 | *Saint Andrew* | n/a |

| No. | As Built | Later Changes |
|---|---|---|
| 2914 | *Saint Augustine* | n/a |
| 2915 | *Saint Bartholomew* | n/a |
| 2916 | *Saint Benedict* | n/a |
| 2917 | *Saint Bernard* | n/a |
| 2918 | *Saint Catherine* | n/a |
| 2919 | *Saint Cecilia* | *Saint Cuthbert* (Oct 07) |
| 2920 | *Saint David* | n/a |
| 2921 | *Saint Dunstan* | n/a |
| 2922 | *Saint Gabriel* | n/a |
| 2923 | *Saint George* | n/a |
| 2924 | *Saint Helena* | n/a |
| 2925 | *Saint Martin* | n/a |
| 2926 | *Saint Nicholas* | n/a |
| 2927 | *Saint Patrick* | n/a |
| 2928 | *Saint Sebastian* | n/a |
| 2929 | *Saint Stephen* | n/a |
| 2930 | *Saint Vincent* | n/a |
| 2931 | *Arlington Court* | n/a |
| 2932 | *Ashton Court* | n/a |
| 2933 | *Bibury Court* | n/a |
| 2934 | *Butleigh Court* | n/a |
| 2935 | *Caynham Court* | n/a |
| 2936 | *Cefntilla Court* | n/a |
| 2937 | *Clevedon Court* | n/a |
| 2938 | *Corsham Court* | n/a |
| 2939 | *Croome Court* | n/a |
| 2940 | *Dorney Court* | n/a |
| 2941 | *Easton Court* | n/a |
| 2942 | *Fawley Court* | n/a |
| 2943 | *Hampton Court* | n/a |
| 2944 | *Highnam Court* | n/a |
| 2945 | *Hillingdon Court* | n/a |
| 2946 | *Langford Court* | n/a |
| 2947 | *Madresfield Court* | n/a |
| 2948 | *Stackpole Court* | n/a |
| 2949 | *Stanford Court* | n/a |
| 2950 | *Taplow Court* | n/a |
| 2951 | *Tawstock Court* | n/a |
| 2952 | *Twineham Court* | n/a |
| 2953 | *Titley Court* | n/a |
| 2954 | *Tockenham Court* | n/a |
| 2955 | *Tortworth Court* | n/a |

## WESTERN TIMES SPECIAL

### Naming Policy

Names carried by early locomotives (numbered between 98 and 190) were remarkable for their diversity. The christening of No. 100 was a gracious tribute to Churchward's predecessor and the naming of No. 171 was an obvious choice for a locomotive engaged in comparative assessment against the French 4-4-2 No. 102 *La France*. As a symbol of progress, fitting of *Vanguard* to No. 98, the first true standard 2-cylinder 4-6-0 was appropriate but the gesture was belated and short-term. It is reported that this locomotive also carried the name *Persimmon* (after the thoroughbred racehorse) for a short period during 1906/ 07. Oddities were the first names applied to Nos. 172/ 9, as these related to defunct stagecoach services that had operated in what later became GWR territory.

A theme emerged early with use of GWR director's names, as carried by Nos. 98 (3rd name)/ 174/ 175 (1st, 2nd & 3rd)/ 176/ 177/ 178 (2nd)/ 184 (1st & 2nd)/ 185 (1st)/ 187 (1st). In the case of No. 178 (1st), recognition was bestowed upon a director's racehorse rather than the man. While it might have been politic to honour certain members of the board, progression from ordinary mortal to knighthood to lord of the realm necessitated more renamings than might have been desirable. It is speculated whether this policy had a perceived discriminatory aspect vis-à-vis those directors condemned to remain plain 'misters'.

More or less contemporaneous with directorships, other names applied had literary connotations. Nos. 171 (2nd)/ 185 (2nd)/ 186/ 188-190 plus No. 2906 were the titles of novels by Sir Walter Scott. Nos. 172 (2nd)/ 179 (2nd)/ 180/ 181/ 183/ 184 (2nd)/ 187 (2nd) recognised characters in his novels. An oddball was No. 182 *Lalla Rookh*, which recognised a character in the eponymous poem by Thomas Moore. Locomotives in the 171-90 number series were collectively known as the 'Scott Class' although this title apparently fell away with standardisation of the 4-6-0 wheel arrangement and formal adoption of the 'Saint' title for the whole 29xx number series.

Although more obviously themed, the titles borne by Nos. 2901-5/ 7-10 had a whimsical flavour. The following were drawn from characters in literature: Nos. 2901 (novel by Eliza F Pollard)/ 2903 (novel by B Lytton)/ 2907 (Shakespeare's 'Much Ado About Nothing')/ 2908 (novel by FH Burnett)/ 2909 (poem by Mrs Hemans). The remainder had colourful backgrounds: Nos. 2902 (of doubtful reputation, mistress of the Arthurian magician Merlin whom she supposedly murdered before marrying Sir Pelleas)/ 2904 (wife of the Earl of Mercia but better known for her horse-riding skills)/ 2905 (instigator in the murder of Scotland's king)/ 2910 (unrequited lover of Sir Lancelot in Arthurian times).

In contrast to the glorious names of the Broad Gauge fleet, emergence of alphabetically-regimented themes was first evident with Nos. 2911-30. A curious change concerned No. 2919 when Saint Cecilia (a 3rd Century martyr) became Saint Cuthbert (a 7th Century Bishop of Lindisfarne). Both were important ecclesiastical figures of their respective eras but the reason for the replacement is unclear.

**Above: Nameplate from No. 2975 *Lord Palmer*. Built as 4-6-0 No. 175 in March 1905, it ran unnamed for two years until being christened *Viscount Churchill* in March 1907 following transfer of that name from No. 184. Renaming followed in February 1924 to *Sir Ernest Palmer* in recognition of the GWR Company Director, altered again to the form shown here upon his elevation to the peerage in October 1933.**

### Series Titles

The number blocks 2901-10, 2911-30, 2931-55 were sometimes colloquially referred to as the 'Lady', 'Saint' and 'Court' series respectively; 'Scott' in this context seems to have been little used. With the exception of No. 2952 all the Courts related to buildings located in GWR territory, and two were linked by ownership to members of the company's hierarchy. *Lord Hillingdon* (No. 2945) was the GWR's auditor and his son (A H Mills) later became a director. *Stackpole Court* (No. 2948) was owned by Earl Cawdor, director. The Courts initiated the Company's anodyne practice of using private addresses of the upper classes.

**Above: No. 2940 *Dorney Court* was a December 1911 build, the final locomotive constructed under Lot. 185. Released to traffic in Great Western lined green livery, which included the obligatory single orange lining on the nameplate riser. The Grade 1 listed Tudor mansion of Dorney Court is situated north-west of Windsor and is regarded as one of the finest manor houses in the whole of England.**

## Livery

During GWR days, the Saints were always rated as express passenger locomotives and as such were adorned with the appropriate livery of the time. Nos. 98 and 100 entered service in time to receive the last version of the Dean era late-Victorian livery i.e. Lined middle chrome green for boiler cladding, firebox, cab sides, tender sides; Indian red for splasher fronts, wheels, frames; Polished steel, brass and copper as appropriate; Lining in orange/ black / orange; Tenders lined in three panels with central entwined monogram. There is also an official photograph of No. 171 in works grey in this livery (Page 13) but as noted the accurate provenance of this view is doubtful.

From 1904, the Indian red was dropped in favour of lined black and a single lining panel on the tender sides enclosing the legend '**GREAT** (garter crest) **WESTERN**'. In the 1920s, the garter crest give way to the 'two cities' coats of arms, then the art deco 'shirt button' emblem from 1934, and finally the plainer unlined wartime black. Post war witnessed efforts to restore green livery and the glory of earlier years with '**G** (twin cities) **W**' on the tender as illustrated below.

With numbers in serious decline, British Railways chose to treat the class as mixed-traffic machines which meant neo-LNWR black lined red, grey and white. This new corporate treatment thankfully failed to diminish the aesthetic allure of Churchward's classic.

**Above: No. 2931 *Arlington Court* stands at Exeter St. Davids with a Down train on 13 September 1947. The Bristol Bath Road (BRD) allocated locomotive still carries the wartime economy plain green livery, but is at least paired with a G (crest) W adorned tender.**

# DESIGN EVOLUTION - BOILER

Within four years of the experimental trio's arrival (Nos. 100, 98 and 171), G J Churchward delivered a paper based on practice as much as on theory to the Institute of Mechanical Engineers which commenced with the following statement:

> *'The modern locomotive question is principally a question of boiler. The great increase in the size of boilers and in the pressures carried, which has taken place during the past few years, has necessitated the reconsideration of principles of design which had been worked out and settled during many years' experience with comparatively small boilers carrying low pressures. The higher temperatures incidental to the higher pressures have required the provision of much more liberal water-spaces and better provision for circulation. Locomotive engineers have now settled down to the use of one of the two types of boiler for very large engines, the wide firebox extending over the frames and wheels and the long narrow box sloping up over the axles behind the main drivers.'*

The essential truth of the great man's words was manifest in the speed with which the modern locomotive fleet was created, comprising the following wheel arrangements: (four cylinder) 4-6-0, 4-4-2, 4-4-2T, 4-4-0, 2-8-0, 2-8-0T, 2-6-0 and 2-6-2T. All followed the core design principles first manifest in the boilers built for the Saints and their 2-8-0 counterparts.

## The Background

The closing years of the 19th Century saw the railway industry generally confront the need for larger locomotives and while differing solutions were tried with varied degrees of success, by 1910 it was evident that the GWR's locomotive pre-eminence was unquestionable.

The change process that took a large railway company reliant on six-wheeled tender and tank locomotives (plus the Dean 4-2-2s) to this position was driven by several factors. The residual broad gauge network had become a costly burden by 1892 and its removal continued to impose expense in follow-up infrastructural work that was largely completed by the turn of the century. Perhaps the improvement with greatest direct impact upon motive power policy was the opening in December 1886 of the Severn Tunnel. Built under arduous conditions between 1873 and 1885, at 4.4 miles long it would remain the world's longest underwater tunnel for over 100 years. Although its construction eliminated the 'Great Way Round' sobriquet and enabled important revision of timetables, the tunnel presented unusual operating challenges due to length, limited ventilation, and especially its gradient profile. Eastward from Severn Tunnel Junction, the line descended at 1-in-90 for 3½ miles to the centre of the tunnel followed by 3½ miles at 1-in-100 up to Pilning, and after a short level stretch another 3½ miles at 1-in-100 as far as Patchway. Speed in the tunnel was limited to 60 mph although it rarely exceeded 30 mph. A locomotive stalling or

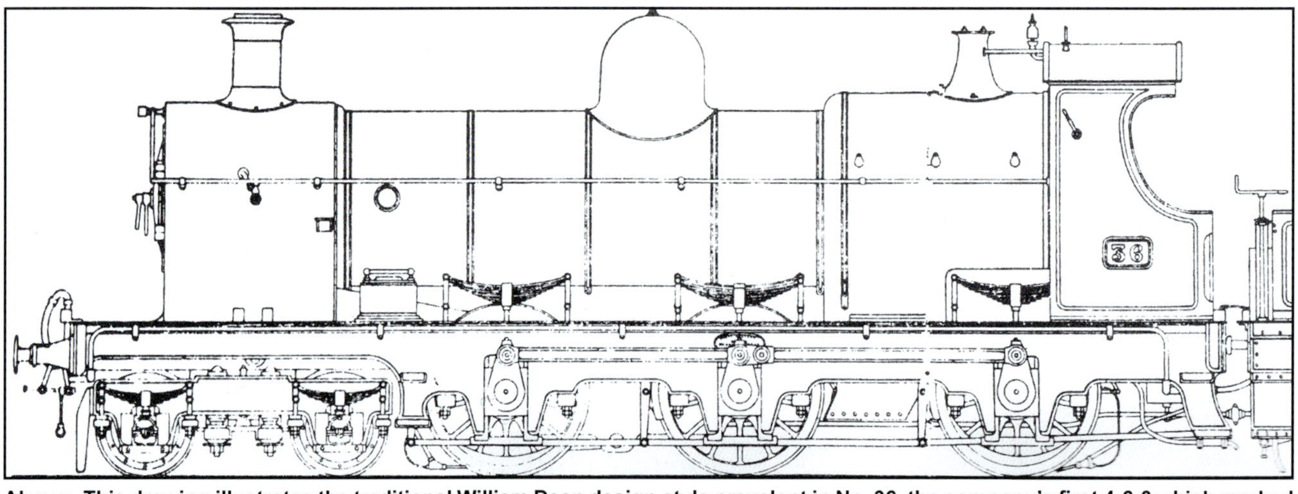

**Above: This drawing illustrates the traditional William Dean design style prevalent in No. 36, the company's first 4-6-0 which marked start of the search for a viable large boiler. This unique locomotive was pure Dean in its low-pitched boiler with raised round-topped firebox, the double-frame chassis structure with inside cylinders, the springs mounted above the running plate, and classic outside-framed bogie. The elegant style and the embellishment of brass, copper and ornate livery typified the GWR's late Victorian years. Invisible but significant was the installation of Serve tubes whose profile increased the heating surface in an attempt to maximise steaming capacity. This locomotive was Dean's largest design and its 'crocodile' nickname was appropriate in view of its length.**

failing over this difficult section would have a serious impact upon heavy traffic volumes.

Double-framed 4-6-0 No. 36, William Dean's last and largest design was conceived specifically to overcome these operating challenges. In establishing what enlargement could offer, it did all that was required by single-handedly traversing the tunnel in 11 minutes with 30 loaded wagons and two vans whereas a pair of 0-6-0s could at best manage a comparable load over the same stretch in 18-20 minutes. No. 36 had the reputation for being 'immensely strong' although this was in comparison with the modestly sized 0-6-0/ 0-6-0STs which then dominated GWR freight services.

The Dean-Churchward transition was probably the most successful inter-generational partnership of the steam saga. Under Dean's wing, Churchward investigated ways and means for coping with the greater loads and faster speeds that the new century would demand. Whatever lasting solution would result, a continuously high level of steam production was paramount, as summarised in his avowed intention to build a locomotive that could generate 2,000 drawbar horsepower at 70 mph. The dimensional variety of locomotive boilers introduced in the last decade or so of the 19th Century was remarkable but key stages can be measured by advances in freight locomotive design:-

| **Class/ No.** | **2301*** | **36** | **2601** |
|---|---|---|---|
| **Built** | 1891-5 | 1896 | 1899 |
| **Wheel Arrangement** | 0-6-0 | 4-6-0 | 4-6-0 |
| **Driving Wheels** | 5' 0" | 4' 6" | 4' 7.5" |
| **Cylinders** | 15.5" x 26" | 20" x 24" | 19" x 28" |
| **Boiler Style** | Front Dome | Centre Dome | Domeless |
| **Boiler Pressure** (lb/sq in) | 150 | 165 | 180 |
| **Boiler Barrel** | 10' 3" x 4' 5" | 14' 0" x 4' 7" | 10' 6" x 4' 10" |
| - pitch | 7' 3" | 7' 9" | 8' 9" |
| **Firebox** | Flush | Raised | Belpaire |
| - length | 5' 4" | 7' 0" | 7' 0" |
| - width | 3' 4" | 5' 10" | 5' 10" |
| - height | 6' 0 | 5' 3" | 4' 9" |
| - combustion chamber | - | - | 3' 6" (length) |
| **Heating Surfaces** (sq/ft) | | | |
| - tubes | 1180 | 1402 | 1713 |
| - firebox | 108 | 116 | 167 § |
| **Grate** (sq/ft) | 17.2 | 30.5 | 32.2 |

* Dean Goods fitted with S2 type boiler.

§ No. 2601's firebox surface included combustion chamber.

The Dean Goods/ No. 36 comparison highlights the size disparity between 19th Century tradition and future expectations while No. 2601 reflected radical thought about how boilers might develop to meet the operating standards that Churchward sought. The sequence of experimental designs reveals some of the exploratory features that came under scrutiny. No. 36 was one of the earliest 4-6-0 types in the UK and the largest design that could be solely attributed to Dean. Entirely in the GWR's late 19th Century tradition, this handsome machine revealed shortcomings. The long boiler induced stresses that led to tube leakage at the throatplate while space limitations constrained rear end efficiency. The typical Dean raised firebox to some extent increased water volume while the large area of the firebox base required termination of the inside frames immediately behind the rear driving wheel axle. Conclusions drawn included:

- The long, thin barrel was poorly proportioned relative to the rear end water spaces.
- Barrel elongation failed to support high steaming rates for sustained periods and induced front end heat loss.
- Enlargement of an essentially outmoded concept was non-viable.
- A totally fresh approach to design was needed.

Another experimental locomotive, 4-4-0PT No. 1490 (built 1898) revived a Broad Gauge era concept of a firebox wider than it was long. The main objective of this unsuccessful project was exploration of a more powerful replacement for the 2-4-0T/ 0-4-2T fleet. Although contemporary designs, No. 1490 differed in carrying an early experimental firebox of the Belpaire form leading to the following key variances (No. 36's dimensions in brackets):- boiler barrel – 10' 8" x 4' 7" (14' 0" x 4'7"); firebox heating surface – 93 sq ft (116 sq ft); grate – 20.4 sq ft (30.5 sq ft). These differences indicated that refinement of boiler and firebox shape was probably more promising than simply making the vessel bigger.

The other significant exploratory exercise of this period commenced with 4-6-0 No. 2601 where the Dean-style chassis format disguised revolutionary features including the long (by contemporary standards) piston stroke of 28", piston valves of 8.5" diameter, single slide bars, and cylinders inclined at 1 in 7. These features were clearly at Churchward's instigation and contributed to his views on future policy e.g. his dislike of inclined cylinders. However, the radical nature of the design was more evident above the running plate with No. 2601's boiler. The barrel was 3' 6" shorter than that on No. 36 while the firebox length and width was broadly comparable with the earlier locomotive. The grate area at 32.2 sq ft represented a modest increase over that of No. 36, and was then the largest in Britain. There was an illusory effect imparted by the higher boiler pitch as the firebox height was actually six inches less than that of the earlier 4-6-0 although the Belpaire form was important in increasing the water volume around the firebox. Use of 4' 7½" diameter driving wheels implied confinement to freight duties but allowed continued use of a wide firebox as emphasised by the vertical sides of the casing. Inclusion of a combustion chamber re-affirmed the determination to extract maximum energy from the fuel. The large saddle sand box mounted on the boiler behind the chimney added to the awkward appearance which was at odds with Dean's earlier trademark elegance. Dislike generated by its ungainly appearance helped it earn the 'Kruger' nickname after public enemy No. 1, leader of the Boers in the contemporary South African conflict.

No. 2601 embraced promising ideas – probably too many at once – with net results that proved disappointing in practice. In face of this reputation, multiplication of the type might have seemed strange but nine more were built. Planned and constructed as a 4-6-0, the bogie of No. 2602 was replaced with a pony truck before it entered service in June 1901, a change made possible due to use of a successful truck based on an American design that had eliminated the need for a bogie with freight locomotives. A further change was provision of laminated springs for the leading coupled axle while retaining volute springs for the other axles (double nests for the driving and single nest for the trailing). Nos. 2603-10 were designed as 2-6-0s, and delivered January to June 1903; it is speculated that multiplication might have been intended to provide test beds for other novel ideas which were rendered unnecessary by other developments. This contention is supported by completion of drawings for a half-cone tapered boiler barrel which would have eliminated the combustion chamber while retaining the original wide firebox, and also the smokebox. This project pre-dated introduction of the first Standard No. 1 half-cone boiler by around six months and its success led to abandonment of the projected Kruger rebuild exercise in early 1905.

Nos. 2603-10 gained a doubtful distinction as among the shortest-lived of all standard gauge GWR engines, with some possibly ceasing work well before their official withdrawal dates. However, the project was probably less costly than individual histories might imply as the boilers were retained for stationary use at Swindon works and mechanical components were recovered for re-use in the 2-6-0 Aberdare construction programme. On a broader canvass, the Kruger family was an extension of William Dean's experimental work which in the spirit of research and development provided valuable empirical evidence to identify ideas worth following and, equally importantly, those best avoided. The Kruger/ Aberdare transition was part of continuum that saw evolution of the Standard Nos. 2 and 4 boilers which were integral to design of medium/ smaller locomotives in the modernisation programme.

### The American Element

By 1900, it was apparent that none of these experimental ventures had confirmed the broad outline of a boiler viable for use in a large express passenger locomotive yet after a remarkably short interval, Swindon was delivering ground-breaking, highly successful 2-cylinder 4-6-0s that would have a lasting impact upon UK steam locomotive practice. A key determinant in this transition was Churchward's open-minded interest in reaching beyond his peer group's embedded parochialism to seek solutions overseas and through this means, the potential of the Belpaire firebox was identified.

Alfred Belpaire's invention of 1864 had failed to gain much traction beyond his native Belgium until the early 1880s. Then in a circuitous manner, a major US manufacturer recognised the potential of the Belpaire principle. Englishman R P C Sanderson became aware of the type through personal contacts in Germany where he had qualified as an engineer. Sanderson was by then working for the Shenandoah Valley Railroad and took the idea to that concern's motive power superintendent who immediately grasped the potential. Orders for freight and passenger locomotives so equipped were duly placed with Grant Locomotive Works, New Jersey, and more significantly with Baldwin of Philadelphia.

The Pennsylvania Railroad (PRR) was a leading Grade 1 operator which worked an extensive network throughout the most densely populated region of the USA (by the 1920s it was responsible for 11,600 route miles). It was one of the comparatively few US companies that built its own locomotives but also maintained a close relationship with Baldwin in a manner reminiscent of the multi-facetted connections between the Great Central and Beyer Peacock at Gorton, Manchester. Starting about 1890, PRR motive power policy became distinctive in its adoption of the Belpaire firebox in contrast to other US operators who continued to favour the round-topped or Wootten varieties. Virtually every 20th Century PRR class was so equipped across its full spectrum of rigid-framed types from 0-4-0 yard switcher to 2-10-2 heavy freight plus the dual-power unit 'Duplex' classes and the solitary 6-8-6 turbine locomotive. Semi-articulated Mallet 2-8-8-0/ 2-8-8-2 classes were also included.

The familiar Belpaire shoulders were prominent but in taking advantage of the more generous loading gauge, the firebox sides typically slanted outward down to the running plate and with several classes, combustion chambers were included. The PRR was probably the greatest exponent of the Belpaire firebox worldwide, and perhaps the best-known class was the redoubtable K4 Pacifics of which 425 were built between 1914 and 1928. These extraordinary machines stayed in front line service until the end of PRR steam, competing successfully against larger locomotives of the New York Central system. They carried Belpaire fireboxes (without combustion chamber) and were married to what Americans refer to as a 'conical boiler'.

Around the turn of the century, UK mechanical engineers tended to disdain American practice for crudely built end-products with shorter operating life expectancies. In a pure quality sense, this view had some cogency but

**Above: The Pennsylvania Railroad was probably the most prominent user of the Belpaire firebox worldwide as exemplified in this view of 2-8-0 PRR Class H6 No. 1499. The locomotive was on display at the Louisiana Purchase Exposition, more commonly known as the 1904 St. Louis World's Fair (held 30 April to 1 December). It was standing on the PRR's stationary testing plant which presumably was brand new and awaiting permanent installation at the company's works in Altoona, PA.** ***Missouri Historical Society.***

took little account of vastly different operating conditions in terms of distances, average journey speeds, and loads. Further, the American solution apparently satisfied cost-benefit expectations while meeting the critical need for reliability. British opinion tended to overlook the contrast between a domestic network where depot-based assistance was rarely more than 50 miles distant, and the vast emptiness of prairie, steppe and bundu.

Records appear opaque on the nature of any contact between Swindon and the PRR during the 19th Century but a Locomotive Committee minute dated 22 January 1902 confirms that J W Cross, Assistant Manager, Locomotive Works had been granted ten weeks paid leave of absence to visit America 'to see what is going on in electricity there'. This could have meant investigation of the use of electric power either in workshop practices or in meeting traction requirements. However, a minute dated 11 June 1902 confirms reimbursement of extra expenses for F C A Coventry, draughtsman, who had accompanied Cross on the US visit.

It is hard to see what purpose a draughtsman could usefully fulfil concerning investigation into 'electricity' but locomotive design practice, especially regarding boilers was another issue entirely. The chronology which is so often an important element in tracing the evolution of steam locomotive design certainly fits. It appears that the cheaper, short life steel firebox on No. 100 was there to prove its operational capacity while 'boxing safe' with a conventional parallel boiler. The second developmental stage was to investigate the degree to which a conical boiler could contribute incrementally to steam raising capacity, and thus justify the additional construction cost. In this respect, so far as the GWR was concerned, the boiler inaugurated with No. 98 little more than a year after No. 100's appearance was less than a pure cone as the bottom edge of the vessel was parallel with the running plate, as would be all following Great Western tapered boilers.

Although the PRR seems to have been the prime target, contact was evidently established with other operators as drawings were supplied in September 1902 by the Lake Shore and Michigan Southern Railroad (LS&MSRR). This material, which described a wide-firebox 2-6-2 used to work the New York-Chicago '20th Century Limited', presumably only over the section of that key artery which lay within LS&MSRR territory.

This stimulated speculation that Swindon might be considering an express locomotive with this wheel arrangement as an alternative to a 4-6-0. With the recent elimination of the need for a leading bogie on freight locomotives, this feature might have been thought possibly feasible for faster machines. It was probably more significant that this operator was a satellite of Cornelius Vanderbilt's empire and later fully absorbed into the New York Central system. Obviously, the GWR's enquiries/ investigations extended beyond the PRR-Baldwin nexus and it is possible that the conical boiler designed for the Krugers (as mentioned previously) derived from these discussions.

Although probably incidental to the wider Saints' story, it is worth noting that American practice continued to sustain the GWR's interest as evidenced in the following Locomotive Committee minutes:

– **26 February 1903**: It was agreed that John Armstrong, the Divisional Superintendent, Locomotive & Carriage Department at Paddington should visit America (duration unspecified) 'to study railway and other practice in that country'. This may have been connected with US-style features in rolling stock design that emerged in the 1900s.

– **12 June 1906**: A report by G H Pearson, Assistant Manager, Locomotive Department was tabled following his recent visit to the USA (duration and purpose of visit unrecorded).

**Above: No. 100 *William Dean* reversing out of Paddington, with modified boiler and modernised tender with side fenders. The original experimental boiler and firebox (as seen on Page 11) were scrapped after about 12 months, and replaced by this D2 half-cone saturated boiler and a standard form copper firebox. The locomotive was withdrawn in June 1932 (by then No. 2900) when its non-standard cylinders had reached their re-boring limit.** ***Stephenson Locomotive Society.***

## The Two-Cylinder 4-6-0s

While the effectiveness of the Belpaire firebox with a wide base for small-wheeled locomotives was proven, the focus with 2-cylinder 4-6-0 No. 100 was on whether the narrower firebox enforced by the 6' 8½" driving wheels could cope with the enhanced demand for steam. Encouragement must have been provided by the superior calorific quality of fine Welsh coal although later experience would show that performance levels were sustained with locomotives allocated to the GWR's northerly depots which often relied on coal emanating from fields other than the Valleys.

However, there was more to boiler superiority than form alone. Churchward was a Devonian whose career had started in the South Devon Railway's workshops at Newton Abbot. In a more confined environment than the departmentalised world of Swindon, he would have encountered in microcosm the challenges presented by a complex engineering enterprise. Further, this intimacy ensured close connection with the operational experiences of the works' products and the men who drove them. This holistic exposure fostered understanding of the practical issues engaged in the abrupt change in operational demands from the racing stretch down the Exe valley to the assault on the southern shoulders of Dartmoor.

In later years while considering the running of non-stop Paddington-Plymouth services, he would have understood the demands upon man and locomotive over that notorious final stretch, having already been hard at it for 214 miles over the 'short way round' via the Berks & Hants route.

An ancillary aspect to effective generation of steam was its preservation. Rather than adopt the more modern 'pop' safety valve that would blow suddenly and instantaneously reduce working pressure before its closure, the traditional spring-loaded Ramsbottom type was retained, housed within the familiar brass bonnet. Careful nurturing allowed the valve continuously to 'sizzle' at or just below 225 lb/ sq in. Churchward maintained warm relationships with subordinates, often entertaining consideration of ideas that might be contrary to his own preconceptions. This leadership style extended beyond the drawing office and embraced the cadre of running inspectors, all ex-drivers of long experience. While riding the footplate, they coached young firemen in how best to make that safety valve sizzle.

Churchward's vision and his consequential development work served the Great Western well. The boiler was used on the Saint class's illustrious contemporaries - the aptly-named 4-cylinder 4-6-0 'Stars' and the splendid 2-cylinder 2-8-0 Class 28xx. A generation later, it was used with the 2-cylinder 4-6-0 classes 'Hall', 'Grange' and 'Modified Hall'. By 1909 the key features of the superheated full-cone Standard No. 1 boiler had been finessed and despite subsequent relatively minor variations in heating surfaces, 928 examples were built between then and 1960. Later increased haulage demands saw enlargements, chronologically:- No. 7 on 2-8-0 Class 47xx in 1922, No. 8 on the Castles in 1923 and No. 12 on the Kings in 1927. On a pound-for-pound comparison, it seems unlikely that any these later additions did their work better.

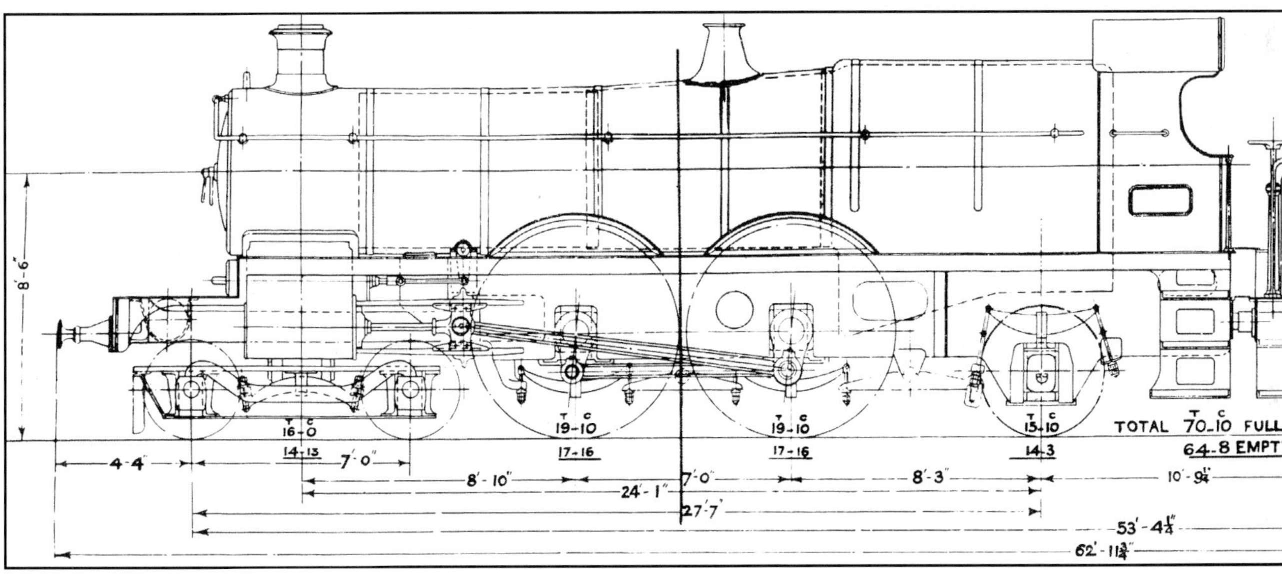


**BOILER TYPE COMPARISON - Above: Saint with 4-4-2 wheel arrangement, half-cone saturated boiler and short smokebox.**
**Below: No. 100 in its final condition with full-cone boiler, top feed and extended smokebox.**

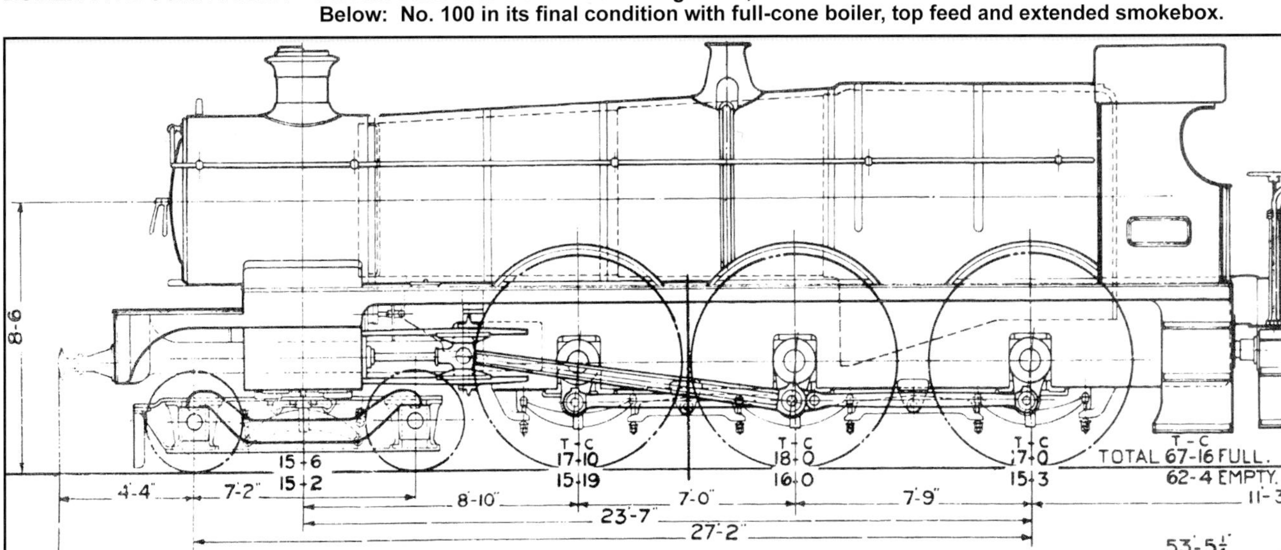

## The Acid Test

A key question arising from boiler design and driving style was whether recovery of the greater marginal build cost attributed to the Belpaire/ tapered barrel combination as compared with products of the 'Doncaster school' could be achieved through superior fuel consumption. Objective non-partisan evidence was therefore important, drawn, for example, from a broader base than mere comparison of two locomotive types as in the 1925 Exchange Trials, or from anecdotal evidence. On 1 June 1938, the Railway Research Service produced (for limited circulation) comparisons of a wide range of statistical data concerning the activities of the Big Four between 1 January 1923 and 31 December 1937. The RRS was an independent body and its detached presentation of informed statistics provided useful material from which to challenge long-cherished beliefs. The report included the following statistics for fuel consumption measured as Average Pounds of Coal Consumption per Engine-mile across all traffic categories:

| Year | GWR | LNER | GWR as % of LNER |
|---|---|---|---|
| **1923** | 46.2 | 58.1 | 79.5 |
| **1929** | 45.2 | 57.3 | 78.9 |
| **1930** | 45 | 56.5 | 79.6 |
| **1931** | 45 | 56.5 | 79.6 |
| **1932** | 44.3 | 55.3 | 80.1 |
| **1933** | 44.3 | 55.2 | 80.3 |
| **1934** | 44.1 | 55.2 | 79.9 |
| **1935** | 44.4 | 55.4 | 80.1 |
| **1936** | 44.4 | 55.8 | 79.6 |
| **1937** | 44 | 56.5 | 77.9 |
| **Average** | 44.69 | 56.2 | 79.6 |

The consistent 20% efficiency advantage imparted by the Swindon philosophy was thereby incontrovertibly proven. Such evidence was taken to account when fledgling British Railways set about designing a new range of standard classes based upon the Big Four's best practices. That virtuous course could be traced back to Churchward's pioneering work with his initial trio of 2-cylinder 4-6-0s, and to his conviction that motive power excellence started with boiler efficiency.

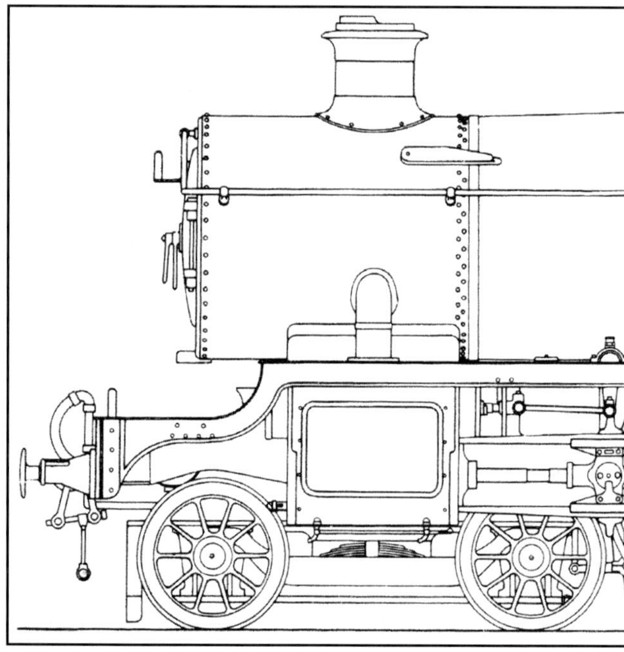

**Above: Side profile drawing of the Saint Class front end in its final form, with outside steam pipes and wide diameter capuchon fitted chimney.**

**Below: Sectioned General Arrangement drawing of the Belpaire firebox as fitted to the GWR Standard No. 1 boiler.**

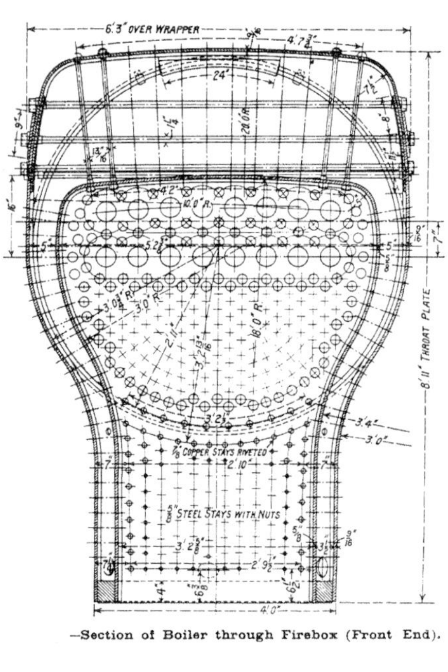

–Section of Boiler through Firebox (Front End).

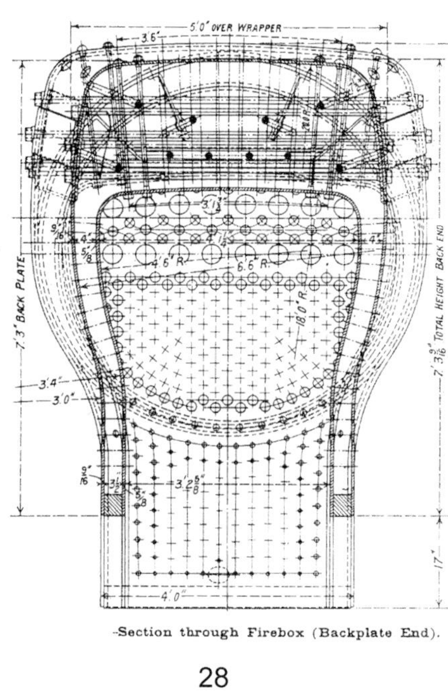

–Section through Firebox (Backplate End).

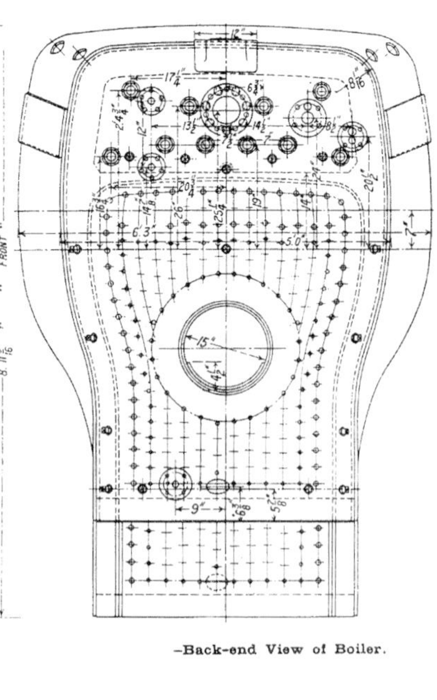

–Back-end View of Boiler.

Above: No. 2938 *Corsham Court* with superheated full-cone boiler, fitted with top feed. The latter device was introduced in 1911 and added to boiler efficiency. Feed water passed through two pipes that rose from running plate level up around the boiler in line with the safety valve and was deposited on perforated trays within the steam space. By the time the water drained through the perforations, its temperature was around 40°F – a simple but effective form of feed water heating. *W H Whitworth.*

Below: No. 2933 *Bibury Court* calls at Maidenhead with a Paddington-Oxford train on 21 June 1952. The locomotive is in the final condition for the Saint Class, featuring outside steam pipes (fitted March 1942) and has even acquired parallel shank Collett buffers. The locomotive spent its final years allocated to Leamington shed (84D), from where it was withdrawn on 19 January 1953. *R C Riley.*

# DESIGN EVOLUTION - CHASSIS

Beyond Churchward's focus upon primary power generation through boiler efficiency lay the question of chassis design. Up until 1893, except for Dean's 'Achilles' Class 4-2-2s, GWR narrow gauge tender locomotives had two-cylinders and six-wheels, confined to essentially simple 2-2-2, 2-4-0 and 0-6-0 designs, depending upon their intended traffic category. Between 1894 and 1902 (the year that No. 100 entered service), growing mainline passenger demands had been satisfied by the 4-4-0 fleet which had expanded from 4 to 205. The numerical peak was achieved at 397 in 1923, and their presence was essential as the company developed its ten-wheeled fleet to meet the exacting standards upon which Churchward had determined at the start of the century.

Single drivers were popular with footplate crews for their smooth running qualities but less so when starting heavy trains or tackling adverse gradients which required better adhesion, as demonstrated by the preference for the 4-4-0 'Duke' Class west of Newton Abbot. Tractive effort was a theoretical standard adopted to vaunt the power of a specific design, often for prestige purposes, but had relatively little practical value. Relating tractive effort to weight borne by the driving axles renders another arbitrary measure in the Factor of Adhesion, which had slightly more practical use in comparing haulage potential. Calculated as the adhesive weight divided by the tractive effort, the optimum under British practice is regarded as '4' with the larger the divergence below this figure indicating the greater propensity for the locomotive to slip under load. The table below charts the comparative data of the key Great Western locomotive designs in traffic around the turn of the century:-

| Built | Class / Loco | Wheel Arrangement | Driving Wheel Diameter | Cylinders | Cylinder Volume (cubic in) | Boiler Pressure (lb/sq in) | Tractive Effort (85%) | Adhesive Weight (tons) | Factor of Adhesion |
|---|---|---|---|---|---|---|---|---|---|
| 1894 | Achilles | 4-2-2 | 7' 8.5" | 19" x 24" | 54,440 | 160 | 12,740 | 18 | 3.2 |
| 1894 | Armstrong | 4-4-0 | 7' 1" | 20" x 26" | 62,080 | 160 | 16,640 | 31.5 | 4.2 |
| 1903 | No. 98 | 4-6-0 | 6' 8.5" | 18" x 30" | 61,080 | 200 | 20,530 | 54.2 | 4.3 |
| 1904 | No. 171 | 4-4-2 § | 6' 8.5" | 18" x 30" | 61,080 | 225 | 23,090 | 39 | 3.8 |

§ As temporarily modified for comparative trials against the de Glehn compound No.102.

A significant aspect of these results concerns the 4-2-2 and the 4-4-0. They shared the same boiler plus close relationship in size, overall weight, and handsome appearance but the 4-4-0 was a sluggish performer, attributable to the greater aggregate cylinder volume. As the design afforded little potential for a larger boiler, the converse interpretation is that the Armstrongs were 'over-cylindered' which underlines the interdependent and holistic nature of factors relevant to successful enlargement. With few scientific evaluation methods available, development largely relied upon trial and error.

### Early Research

Without a UK-based precedent to follow, it seems probable that United States practice was investigated for ideas regarding chassis design. Twentieth century American locomotives typically conjure visions of enormity but around the year 1900, there was little size difference between many US designs and larger UK examples. Reverting to the Pennsylvania Railroad, comparison with that operator's 2-8-0 Class H6 family built 1899-1913 and the most modern British consolidation is revealing:-

| Operator | Pennsylvania Railroad | GWR |
|---|---|---|
| **Loco Class/ No.** | H6a | 97 (28xx) |
| **Wheel Arrangement** | 2-8-0 | 2-8-0 |
| **Built** | 1901-5 | 1903 |
| **Driving Wheels** | 4' 8" | 4' 7.5" |
| **Cylinders** | 22" x 28" | 18" x 30" |
| **Boiler Type** | Tapered | Half-cone Taper |
| **Firebox Style** | Belpaire | Belpaire |
| **Boiler Pressure** (lb/ sq in) | 195 | 200 |
| **Heating Surfaces** (sq ft) | | |
| - tubes | n/a | 1989 |
| - firebox | 'wide' | 154 |
| **Grate** (sq ft) | 49 | 27.2 |
| **Weight** (long tons) | | |
| - locomotive only | 91.4 | 68.3 |
| - adhesive | 80.4 | 61.5 |

At 1,707 examples, Class H6 became the PRR's most numerous class and the second most prolific consolidation type in the USA. Later in their careers a number were sold to Chinese operators and the last class member was withdrawn by the PRR in 1958. The weight difference over GWR No. 97 was probably due mainly to the enormous firebox necessitated by the comparatively poor quality of indigenous coal.

Like the GWR, American operators during this period were concerned with the operational viability of larger engines. An emerging problem was 'track pound' technically defined as 'dynamic augment', and in the UK known as 'hammer blow'. The impact was exponential with the increased speeds, heavier axle loadings, and greater reciprocating mass implicit with evolving operating demands. The more progressive elements of the US mechanical engineering fraternity were, like Churchward, unconstrained by parochial limitations and sought inspiration beyond their home shores.

These investigations led to recognition of the power potential, operating economy and above all, the excellent riding characteristics of the 4-cylinder de Glehn compound. PRR purchased an example which became their 4-4-2 No. 2512. In 1904 after shake-down running on the road, it was demonstrated on their stationary testing plant exhibit at the St. Louis World's Fair, apparently following presentation of 2-8-0 PRR Class H6 No. 1499 on this facility (see previous chapter). PRR's investment in the de Glehn compound was perhaps courageous under the nose of the formidable Samuel Vauclain, Director and later Chairman of Baldwin Locomotive Works. He had invented the eponymous compound system which actually failed to solve the dynamic augment dilemma. Other initiatives were then under consideration and the most profound was the Baltimore & Ohio RR's acquisition of America's first compound (i.e. 'classic') Mallet in the same year. In the context of the GWR, the harmony of enquiry and development alongside the PRR (conical boilers, Belpaire fireboxes, the stationary testing plant at Altoona followed by that at Swindon, experiments with de Glehn compounds) was remarkable.

**Chassis Development**

**No. 100 (later No. 2900)**
**Built February 1902 under Lot No. 132**

While the main intent with this experimental locomotive was pursuit of boiler efficiency, investigation also commenced into how best the increased energy could be applied. The chassis was of traditional plate frame construction with inside-mounted valves and motion. The cylinder dimensions of 18" x 30" became the base standard for 2-cylinder 4-6-0s (diameter increased by ½" in later variants) and the trial use of double-ported 6½" diameter piston valves dimensionally broached fresh territory. Later, 7" and 7½" diameter double-ported valves were tried before Churchward opted for the simpler standard of single-port 10" diameter piston valves for the 29xx production series.

The parallel boiler was replaced after little more than a year and later iterations followed the developmental course of the No .1 boiler but the unique cylinder/ valve configuration was retained throughout. Withdrawal of life-expired Saints commenced in late 1931 with Nos. 2909/ 10. When the following year, it was deemed that No. 2900's unique cylinders had reached their re-boring limit, the only viable course was retirement. Nonetheless an active working life of 30 years was a notable achievement for an essentially experimental machine.

**No. 98 (later No. 2998)**
**Built March 1903 under Lot No. 138**

This locomotive, regarded as the first Saint proper, was designed and constructed in parallel with the prototype 2-8-0 No. 97 (later No. 2800 - Lot 139 of June 1903) with which it shared many features, including frame design which, probably influenced by US practice, was an unusual combination of plate and bar form. From the locomotive's rear end forward to immediately behind the cylinders, conventional plate frames were used to which were bolted bar extensions. This was necessary

**Left: The de Glehn compound system with its smooth-running qualities and reduced dynamic augment also excited the interest of American engineers. To explore the potential further, the Pennsylvania Railroad purchased a 4-4-2 compound (their No. 2512 pictured here) from the same manufacturer, more or less concurrent with the GWR's acquisition of No. 102 *La France*. Multi-cylinder rigid frame locomotives were generally unpopular in the United States and so far as can be determined, this exercise was a one-off without lasting impact upon USA design philosophies.**

because the cylinder blocks which incorporated 10" diameter piston valves and the smokebox saddle were cast in two halves and bolted together back-to-back along the locomotive's centre line. Such was the driving cylinder alignment that the co-joined cylinder blocks had to be mounted upon bar frames which at their rear overlapped the front end of the plate frames to which they were in turn bolted. The smokebox saddle was an integral part of this structure.

The resultant cylinder assembly was strong, corrosion-free and immune from risk of steam leakage that was inherent in joints between smokebox and cylinders in more conventional structures. Normally the thrust exerted by the motion was most acute around the cylinder area but the strength imparted by this solid formation adequately absorbed stresses. Further, development of uneven wear in the bores over time could be mitigated by reversal of the co-joined block to thus extend cylinder life.

A possible disadvantage lay in the inherent weakness of the fabricated frame structure which threatened disproportionate damage in the event of even a moderate front-end collision. With the Saints, the 6' 8½" driving wheel diameter allowed the horizontal bar frames to clear the cylinder centre line but as a precaution, a rigidly-braced vertical stay plate was installed in line with the slide bar brackets. This extended upward close to but just short of touching the underside of the boiler which meant that with any unusual shock upon the buffer beam, the stay plate would move upward and the weight of the boiler would absorb the thrust. This archetypal Churchward frame format that was applied with all 2-cylinder 4-6-0s until early 1944 when Hawksworth introduced plate frames throughout, with a different design of cylinder blocks and smokebox saddles. This revision was introduced with 'Modified Hall' No. 6959 in response to contemporary need for manufacturing cost reduction and shorter construction times.

**Above: The three 4-4-2 de Glehn compounds purchased from Société Alsacienne des Construction Méchaniques, Belfort played an influential role in adoption of 4-cylinder propulsion and divided drive as an integral part of GWR locomotive development. Dean had experimented with compound steam in the 19th Century but these were the last of the genre to be used on the GWR. Interest in the concept was briefly revived by F W Hawksworth and W A Stanier, who collaborated in the design for a compound 'Castle' in the late 1920s. A gifted draughtsman also schemed out how a King's performance might be improved using techniques devised by Chapelon in the 1930s. Both proposals received short shrift from Collett. This photograph of the third Frenchman No. 104 depicts it ready for service with GWR fittings, tender and livery but before fitting of the name *Alliance,* which took place in 1907.**

### The French Influence

Concurrent with monitoring the performances of prototype 4-6-0s Nos. 100 and 98, Churchward explored the possibilities offered by the French-built de Glehn compound type through purchase of 4-4-2 (works No. 5409 under Lot No. 146) from Société Alsacienne des Construction Méchaniques, Belfort Works (who also supplied the PRR locomotive). Delivered to the GWR, in October 1903 this locomotive (numbered 102 and appropriately named *La France*) was similar to those already in service with the Nord Railway except for certain GWR features (principally brakes, chimney and tender). It introduced four-cylinder propulsion and divided drive whereby the outside high pressure cylinders relied on slide valves and those inside on low pressure flat valves; individual Walschaerts valve gear sets were provided for each cylinder.

The drive configuration was more complex than anything previously seen on the GWR and contrasted with the simplicity of 4-6-0 No. 100. Investment in so exotic a piece of machinery required Churchward to seek specific Board approval rather than through the usual Locomotive Committee process. Perhaps parochialism, xenophobia, or traditionalism were prerequisites for elevation to the Board; perhaps Churchward strengthened his case by assertion that the PRR across the pond was pursuing similar enquiries. In any event, with purchase of larger de Glehn Atlantics (Nos. 103/ 104 later named respectively *President* and *Alliance*) in May 1905, the trio played an important role in learning about design alternatives, in finessing Swindon's engineering standards, and in guiding the next stage of the epoch, the 4-cylinder 4-6-0 'Star' Class.

No. 102 was delivered in October 1903 and after proving trials, started regular passenger work in February 1904. Determination of how this sophisticated locomotive compared with the home-built product was key to decisions concerning future practice so the third 2-cylinder passenger 4-6-0, unnamed No. 171, was selected for comparative trials. To help objectivity, the Swindon product had its trailing driving wheels replaced with outside bearings, a carrying axle and 4' 1½" wheels in October 1904. Reflecting the international flavour of the contest, No. 171 was named *Albion*. Dimensional comparison between the two locomotives:

| **Locomotive** | **102 *La France* *** | **171 *Albion*** |
|---|---|---|
| **Cylinders** | | |
| - high pressure | 13.38" x 25.19" | 18" x 30" |
| - low pressure | 22.06" x 25.19" | - |
| **Boiler** | | |
| - barrel | 13' 7.5" | 14' 10" |
| - diameter | 4' 9.3" to 5' 0" | 4' 10.75" to 5' 6" |
| - pitch | 8' 3.25" | 8' 6" |
| **Boiler Pressure** (lb/sq in) | 227 | 225 |
| **Firebox Length** | 9' 10" | 9' 0" |
| **Heating Surfaces** (sq/ft) | | |
| - tubes | 2288.8 | 1988.7 |
| - firebox | 167 | 154.4 |
| **Grate** (sq/ft) | 29.5 | 27.1 |
| **Wheel Diameter** | | |
| - bogie | 2' 11.875" | 3' 2" |
| - driving | 6' 8.5" | 6' 8.5" |
| - trailing | 4' 7.875" | 4' 1.5" |
| **Wheelbase** | 6' 10.7" + 6' 0" + 7' 0.6" + 7' 10.5" | 7' 0"+ 5' 4" + 7' 0" + 8' 3" |
| **Axle Load** (tons) | | |
| - bogie | 17.5 | 16 |
| - leading driver | 16.5 | 19.5 |
| - trailing driver | 16.85 | 19.5 |
| - trailer | 13.8 | 15.5 |
| **Total** | 64.65 | 70.5 |
| **Tractive Effort** (85%) lb | 23,710 | 23,090 |
| **Adhesion Factor** | 3.15 | 3.78 |

* Dimensions after metric conversion.

Tests were conducted over an extended period and No. 171 worked as an Atlantic until reversion to 4-6-0 in July 1907. The table above shows the dimensional similarity between the two types, the most significant element being No. 171's greater coupled axle loadings and hence superior adhesion factor. Presumably the duration of the trials would have allowed accurate assessment of maintenance costs which in view of the more complex mechanical layout were probably higher for the Frenchman.

Although two more (larger) de Glehn compounds from the same manufacturer were delivered during 1905, their presence had no impact upon the outcome of the *La France*/ *Albion* contest. The core conclusion drawn from the trials was that the chassis format inaugurated with 4-6-0 No. 98 was the preferred way forward.

### Cylinder and Valve Dimensions

While No. 100 was regarded as an experimental exercise, release of dimensional details concerning No. 98 certainly started tongues wagging, focussed upon the then extraordinary 30" piston stroke, when 26" was generally regarded as the feasible limit. This feature tended to avert attention from other unprecedented dimensions:

- Valve Travel: 5 $^{7}/_{8}$" in full gear.
- Steam Lap: 1 $^{5}/_{8}$"
- Lead: $^{1}/_{8}$" negative in full gear that increased to $^{1}/_{8}$" positive at 25% cut-off
- Piston Valves: 10' diameter (as built, this dimension had started out at 6½" with No. 100).

This iconoclastic combination was integral to achievement of Churchward's targeted 2,000 drawbar horsepower at 70 mph. By commencing with careful investigation of boiler design, he had developed a vessel that capable of generating unprecedented steam volumes. Valves and cylinders had to broach fresh dimensional boundaries to put that greater energy to good use.

In drawing up the design parameters, slide valves had been swiftly discarded as inefficient in contributing to excessive wire-drawing so piston valves became obligatory and thus the need for adequate lubrication. Established engineers universally opined that long-lead, long-travel valves were impracticable because of the susceptibility to excessive wear. This view failed to take account of Churchward's holistic approach and the need for proportionality. The need for a step change in lubrication had already been addressed through the GWR's hydrostatic sight feed lubricator, coupled with the most advanced oils then commercially available.

### Cylinder Alignment

No. 98 was built concurrent with prototype 2-8-0 No. 97, which led to a design curiosity in the early class members. There had been concern that tyre wear on the Consolidation's 4' 7½" driving wheels could lead to the low-slung 18" diameter cylinders fouling the loading gauge at about 18" above rail level. Inclination would have avoided the problem but this measure was anathema to Churchward so the cylinders were mounted horizontally with their centre line 2½" above that of the coupled axles. For sake of uniformity, this off-set was also applied to the early Saints until it was realised that with the larger driving wheels this precaution was unnecessary, so cylinder and axle centres were set on the same plane with later engines. The latest Saint to be built (No. 2999) is true to the early class standards as is readily apparent to the naked eye.

### Two Versus Four Cylinders

A factor that had excited American interest in the de Glehn Compound was its smooth running qualities, imparted by four-cylinder divided drive and this element probably had direct impact upon the Stars. Prototype No. 40 (later No. 4000) entered service in April 1906 as a 4-4-2 for comparison against the larger Frenchmen. Great symbolic importance was attached to locomotive names during this period. Just as No. 171 *Albion* evoked sentiments of national pride, after a few months No. 40 was christened *North Star*, a name redolent of GWR locomotive ascendancy from the early days. As Saints and Stars shared the same pattern of boiler evolution, comparison rested on the differing chassis format.

No. 40 inaugurated the classic cylinder layout of divided drive with the inside pair placed ahead, connected to the leading driving axle while the outside cylinders were connected to the trailing (centre with later 4-6-0s) drivers. Dr W A Tuplin from a partisan standpoint asserted that only one UK operator successfully overcame the design challenges presented by the 4-cylinder 4-6-0, as exemplified by the Stars plus later Castles and Kings. Certainly, the experience with multi-cylindered designs elsewhere was chequered. Overseas, locomotive enlargement resulted in heavier individual motion components which greatly complicated service procedures where mechanical parts were hidden between the driving wheels.

This challenge was later addressed by Gresley with his 3-cylinder conjugated valve gear which minimised the inside reciprocating parts. Promotional ballyhoo certainly stimulated international interest:-

| Operator | Built | Class | W. A. | No. | Wdn. |
|---|---|---|---|---|---|
| Union Pacific RR | 1926 | 9000 | 4-12-2 | 88 | 1956 |
| South African Rlys | 1927 | 18 | 2-10-2 | 2 | 1951 |
| Victorian Railways | 1928 | S | 4-6-2 | 4 | 1954 |
| N.S.W. Govt | 1929 | D-57 | 4-8-2 | 25 | 1962 |
| New Zealand Rlys | 1929 | G § | 4-6-2+2-6-4 | 3 | 1937 |
| New Zealand Rlys | 1937 | G | 4-6-2 | 6 | 1956 |

§ The first NZR Class G was problematic; the chassis were recycled into 3-cylinder 4-6-2s with specially-built boilers; the conjugated valve gear was retained, but widely loathed.

Retention of proven, rugged Stephenson's Link Motion and simple expansion steam cycle with the production run of Saints was a key decision. This universally popular system yielded several advantages including the ability easily to adjust the cut-off setting in small increments for performance optimisation thereby aiding varying demands upon the boiler. It was normal practice to start from rest with a long cut-off i.e. closure of the inlet valve at 70-80% of the piston's power stroke. As the locomotive gathered momentum, the point of cut-off was shortened ('notching up') thus relying increasingly upon steam's expansive property to power the remainder of the stroke, thereby reducing steam and hence fuel consumption.

Admission of steam before the piston arrives at the end of the stroke (the 'lead') provides a cushion at commencement of the next power stroke. An intrinsic advantage with Stephenson Link Motion in conventional form with open eccentric rods (as opposed to the closed eccentric rod set-up) was that it allowed variable lead. This minimises compression and back pressure at low speed in full gear but with increased speed and shortened cut-off, the lead automatically advances and so enlarges the compression cushion. Churchward felt that leads elsewhere were often excessive and wasted steam so he chose a break with tradition. He set the lead so that it was negative in full gear thereby imparting greater 'punch' at commencement of the power stroke when starting, becoming positive as the engine was notched up.

By comparison, Walschaerts valve gear imparts constant lead regardless of the cut-off position which can adversely impact performance under varying operating conditions. This was cogently demonstrated by the Beyer Peacock-built 4-6-4Ts of the Belfast & County Down Railway (BCDR) in Northern Ireland. Gorton Foundry had previously built competent 4-6-4Ts for essentially similar work with the New South Wales Government Railway but Crossthwait, the BCDR's opinionated locomotive superintendent ,thought he knew better. A tortuous, difficult design and construction saga ensued due to his meddling and the BCDR acquired engines with Walschaerts valve gear and constant negative lead that yielded poor acceleration, sluggish overall performance and high fuel consumption in the range of 70-80 lbs per mile. Against this extreme example, on the GWR when hill climbing at fairly long cut-off, the Saints consistently had the edge over Walschaerts-equipped 4-cylinder 4-6-0s.

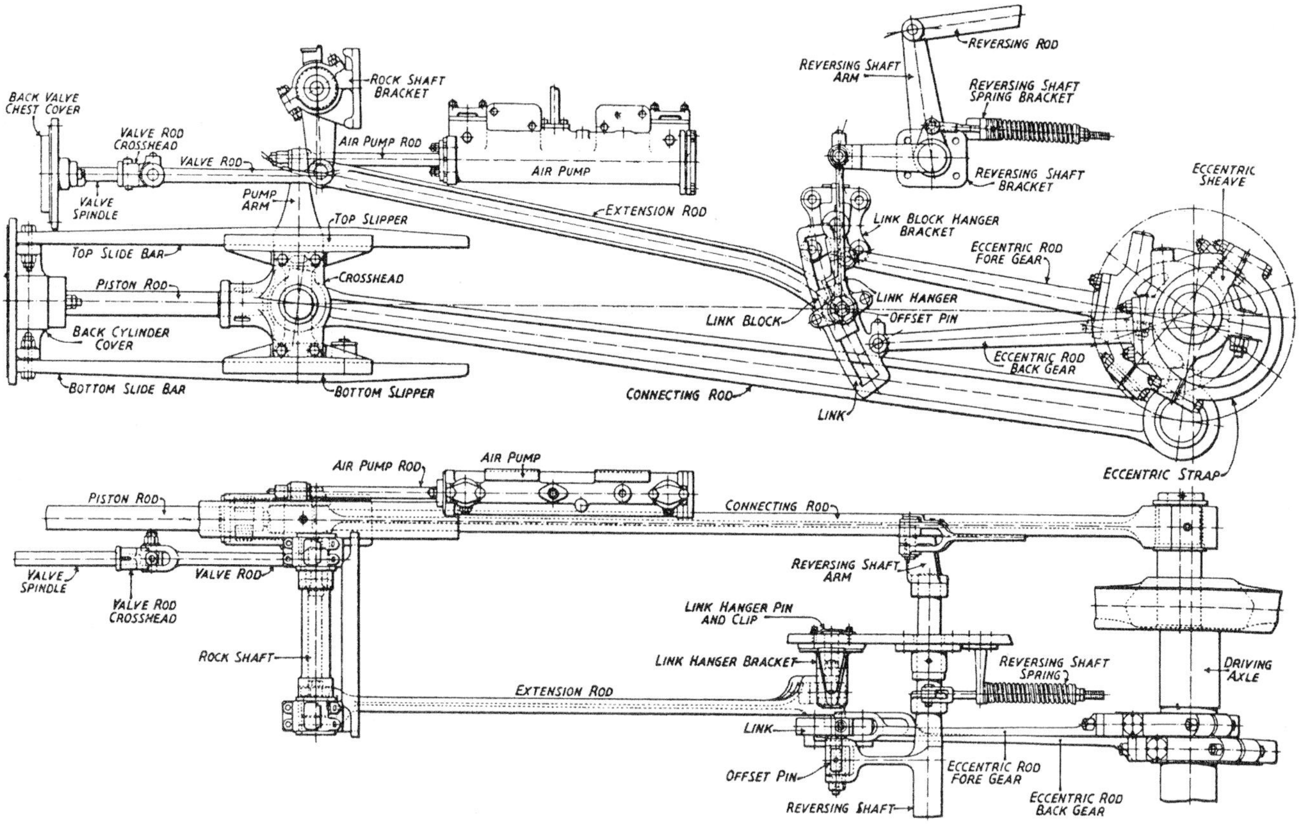


**Above: Stephenson's Link Motion was probably the most popular form of valve gear in the UK in the 19th Century and well into the next. This side view and plan described the layout as would be used in a locomotive with two inside cylinders.**

**Below: With the decision to build 4-6-0s with two outside cylinders, the GWR had to modify the layout as depicted here. This drawing also illustrates the join between the plate main frames which terminated in front of the leading driving axle and the bar frames upon which were mounted the stout back-to-back integrated cylinder blocks. The plate/ bar frame joint was probably the area of maximum stress induced by the fore and aft cylinder motion but no particular problems were encountered as the entire fabricated structure was bolted to the cylinder blocks.**

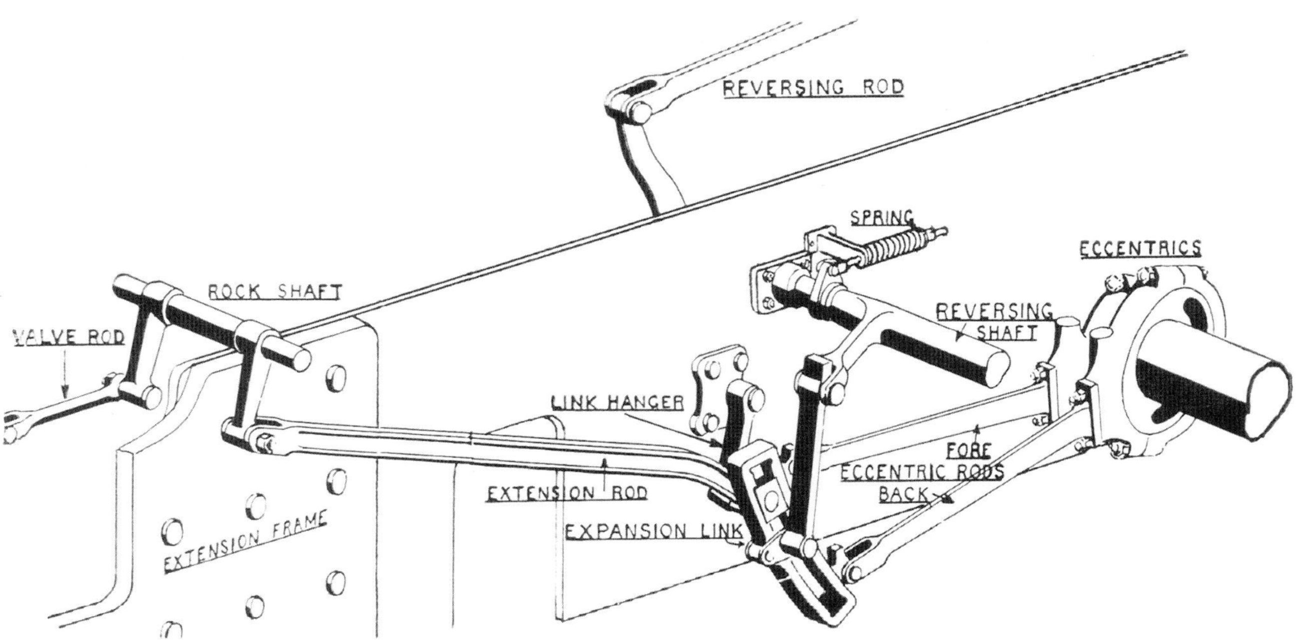

Right: In an attempt to alleviate the gaunt angular appearance of the early engines, 4-4-2 No. 187 was fitted with a fabricated curved front fall plate, although whether during construction or later is unknown. This engine emerged from Swindon as *Robertson* in August 1905 and was renamed *Bride of Lammermoor* in April 1907. Apparently, this was the only example so treated and the addition apparently only served an aesthetic purpose. On conversion to 4-6-0 in June 1912, a curved (concave) fall plate was installed with matching styling below the cab. The location is Bristol Temple Meads sometime during 1908 on what is likely to be an express from the West Country, when it was based at Exeter. *F W Brewer.*

Above: No. 2907 *Lady Disdain* was built May 1906 and was withdrawn in May 1933 in unchanged condition, except for provision of top feed. All class members of 4-6-0 wheel arrangement up to No. 2910 (built May 1906) were delivered in this form i.e. vertical front fall plate with right-angled corners plus straight running plate and prominent cab footsteps. Also, they were equipped with lever (pole) reverse as evidenced by the horizontal reversing arm outside the line of the splashers. The intriguing name is thought to have come from the name addressed to Beatrice by Benedict in *Much Ado About Nothing* by William Shakespeare, although it might have earlier origins. *The Transport Treasury.*

Above: At Old Oak Common on an unknown date (but after fitting with ATC), No. 2988 *Rob Roy* is in its final condition, which was retained until withdrawal in May 1948. This engine had been built as an Atlantic in August 1905, was superheated in August 1910, converted to 4-6-0 in May 1912, and acquired a full cone boiler in May 1914. Despite the significant change of wheel arrangement, it retained lever reverse throughout its career as did the other 4-4-2/ 4-6-0 conversions. *Stephenson Locomotive Society.*

Below: There were instances of experimental modifications to individual locomotives as with No. 2975 *Lord Palmer* (previously *Sir Ernest Palmer* until October 1933) which is seen at Swindon fitted with extended valve spindle guides, as evidenced by the cylinder covers extending forward from the vertical fall plate. The fitting date is unrecorded. *Stephenson Locomotive Society.*

Above: On Saturday 11 October 1947, No. 2906 *Lady of Lynn* had arrived at Birmingham Snow Hill with a terminating express from Cardiff via Hereford, and having run round is about to remove the empty stock. This locomotive was fitted with new cylinders in October 1935 which process included installation of curved front fall plate and valences, plus outside steam pipes. Despite these changes, the lever reverse was retained (the last main-line locomotive so fitted). Between November 1930 and December 1943 Nos. 2902/ 3/ 5/ 8 also received this treatment. No. 2906 was the last to survive in this hybrid state and by the time of its withdrawal, it was something of a 'pet' at its home depot, Cardiff Canton. *P J Garland Collection.*

No. 2937 *Clevedon Court* was built in December 1911 with a superheated full-cone boiler which type it carried throughout its life. New cylinders with outside steam pipes were installed as late as June 1948. It is seen at Hereford in September 1952 in BR lined black livery, its home depot for several years until withdrawal in June 1953. *Lens of Sutton.*

## The Achilles' Heel

By any yardstick, the Saints in their modernity and all-round competence formed the outstanding express passenger class of the United Kingdom in the early Edwardian era, prior to their companion Stars assuming that supremacy. They provided the design format for all 2-cylinder 4-6-0 classes for over forty years using the splendid No. 1 boiler (with the exception of the twenty Manors). This was an extraordinary legacy, typical of the design and construction continuum that was a dominant feature of Great Western locomotive practice.

However, construction of Saints ceased in 1913 whereas new Stars were delivered until 1923 when in the same year, the 'Enlarged Star' appeared with improved cab and larger boiler. Soon known as the 'Castle Class' its production would continue until 1950. Nevertheless, despite the 4-cylinder incursion in the early 1920s, there were reports in the *Railway Magazine* of exciting work by Saints on Paddington-Bristol express services with speeds reaching 80 mph. This work included from 1923 their regular use on a Swindon-Paddington trains including one with a featherweight load of seven coaches (say 230 tons) that was timed to cover the 77.3 miles in 75 minutes start-to-stop (average speed 61.8 mph). They also handled London-Birmingham 2-hour services where their hill-climbing capabilities were much appreciated. Such performances were probably impressive all round, except for the Civil Engineer whose first concern like all of his breed was the care of the railway's physical infrastructure. Hammer blow was the Saints' Achilles' heel and as the superb 4-6-0 fleet multiplied, Churchward's pioneering 2-cylinder 4-6-0s were progressively relegated to second string mainline duties where lower speeds reduced the risk of damage to the permanent way. Thus, Churchward's first 4-6-0 masterpiece graciously yielded pride of place to his second.

Construction of Saints and Stars by year:-

| Year | Saint | Star | Year | Saint | Star |
|---|---|---|---|---|---|
| 1902 | 1 | | 1911 | 10 | 3 |
| 1903 | 2 | | 1912 | 10 | |
| 1905 | 19 | | 1913 | 5 | 5 |
| 1906 | 10 | 1 | 1914 | | 15 |
| 1907 | 20 | 10 | 1922 | | 6 |
| 1908 | | 10 | 1923 | | 6 |
| 1909 | | 10 | | | |
| 1910 | | 7 | **Total** | **77** | **73** |

Dimensional comparison (as built):-

| **Locomotive No.** | **98 (2998)** | **4001** |
|---|---|---|
| **Cylinders** | | |
| - dimensions | [x2] 18" x 30" | [x4] 14.25" x 26" |
| - agg. volume (cu ins) | 61,070 | 66,350 |
| **Wheel Diameter** | | |
| - bogie | 3' 2" | 3' 2" |
| - driving | 6' 8.5" | 6' 8.5" |
| **Wheelbase** | 7' 0" + 5' 4" + 7' 0" + 8' 3" | 7' 0" + 5' 6" + 7' 0" + 7' 9" |
| **Axle Loading** (tons) | | |
| - bogie | 16 | 20.2 |
| - leading driver | 19.5 | 18.6 |
| - centre driver | 19.5 | 18.6 |
| - trailing driver | 15.5 | 18.2 |
| **Total** | 70.5 | 75.6 |
| **Adhesive Weight** | 54.5 | 55.4 |
| **Tractive Effort** (85%) lb | 23,090 | 25,090 |

**Right: Churchward was apparently unimpressed with the French 4-4-2's use of a set of Walschaerts valve gear for each cylinder with the allusion to 'watch-making'. For comparison with the two larger Frenchmen, 4-cylinder 4-4-2 No. 40 (later named *North Star* and renumbered 4000 in December 1912) and was converted to 4-6-0 in November 1909. It was built with specially designed inside 'scissors' valve motion with rocking arms to actuate the outside valves. The scissors design proved very effective and construction commenced on the first ten 4-6-0 Stars with the intention to use this arrangement. However, as No. 40's mileage accumulated, it became apparent that the scissors motion was unduly time-consuming to maintain so the production series used inside Walschaerts motion with outside valves actuated by rocking arms. Thus, almost by accident the classic GWR 4-cylinder format was adopted although it is believed that had time permitted, outside valve gear only would have been preferred i.e. the reversal of the Star/ Castle/ King layout. No. 4000 retained its unique valve motion until it was rebuilt as a Castle in November 1929.**

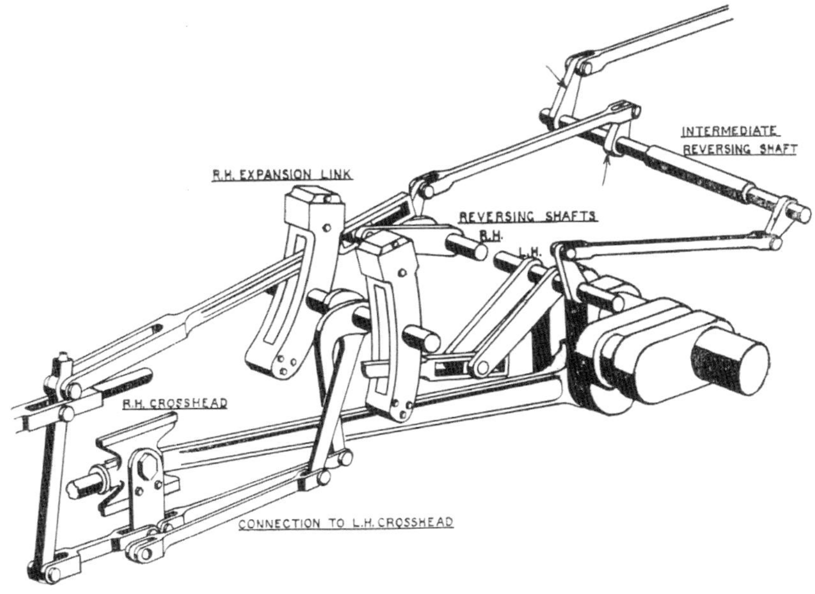

# DESIGN EVOLUTION - SUPERHEATING

Locomotive development in the later stages of the 19th Century focussed significantly upon compound steam to improve thermal efficiency, reduce fuel consumption, and increase power output. Use of the latent energy in exhausted steam to drive a second, larger diameter cylinder(s) had proved successful in steam ships and electricity generation. William Dean had experimented with double expansion steam without realising significant advantage, while Churchward investigated the De Glehn system which was achieving impressive results with Chemins de fer du Nord, France. The purchase of the three 4-4-2s built by Société Alsaciennes des Constructions Méchaniques, Belfort was instrumental in reaching empirically-based conclusions that the advantages gained were insufficient to justify the greater mechanical and hence more costly complexity.

Cylinder dimensions for the GWR's three 'Frenchmen' Nos. 102-4 compared with No. 2901 (the first superheated 4-6-0) and Nos. 4001-10 (first production series of the Star Class):

| Loco Nos. | High Pressure Cylinders | Low Pressure Cylinders | H.P. Cylinder Volume (cubic inches) | L.P. Cylinder Volume (cubic inches) |
|---|---|---|---|---|
| **102** | (x2) 13.38" x 25.19" | (x2) 22.06" x 25.19" | 28,340 | 77,020 |
| **103/ 4** | (x2) 14.19" x 25.19" | (x2) 23.63" x 25.19" | 31,870 | 88,380 |
| **2901** | (x2) 18" x 30" | - | 61,070 | - |
| **4001-10** | (x4) 14.25" x 26" | - | 66,350 | - |

This provides a crude comparison to illustrate the potential for fuel savings with the compound arrangement as the aggregate HP cylinder volume is around 50% or less than that of the simple 4-6-0s with the LP cylinders accounting for the remaining requisite work.

Whereas compound steam re-cycles energy at the primary exhaust stage, superheating increased energy generation by secondary employment of firebox heat before steam reaches the cylinder. Attempts were made to exploit this idea without marked success until modern superheating emerged from experimental work by Wilhelm Schmidt in the late nineteenth century. A Prussian State Railways 4-4-0 Class S3 (of which 1,027 examples were in service) became the world's first superheated locomotive in 1898 and after teething issues had been solved, the equipment proved sufficiently effective for all 104 members of the following PSR 4-4-0 Class S4 to be so fitted in the early 1900s.

Through his interest in foreign motive power development, Churchward took note of the Prussian experience and in May 1906 installed a Schmidt superheater in Saint Class No. 2901 (later named *Lady Superior*) during the locomotive's construction. This was the first British locomotive to be so treated and was applied in a boiler pressed to 200 lb/ sq in. This was soon increased to 225 lb/ sq which became the GWR's standard, a significant change as a contrary view in other quarters held that the presence of superheating could justify a lower working boiler pressure than would be normal with saturated steam.

No. 2901's performance was closely monitored and on 29 June 1906, Churchward reported to the Board *'... although it has been running too short a time to pronounce definitely upon the results, the economy shown up to the present is remarkable and is far beyond any results which have been anywhere obtained by the compound principle.'* He then outlined problems that might emerge over time (burning out of superheater tubes and the possible effect of high temperature steam upon the tightness of valves and glands) and then concluded *'...but even in the face of these drawbacks the results are such that we must certainly pursue the experiment at once'.*

In this succinct statement, he dispelled the proposition that compound steam might provide worthwhile competitive advantage. It was significant that in alluding to possible future problems with superheat, he was confident that the impact of possible negative factors

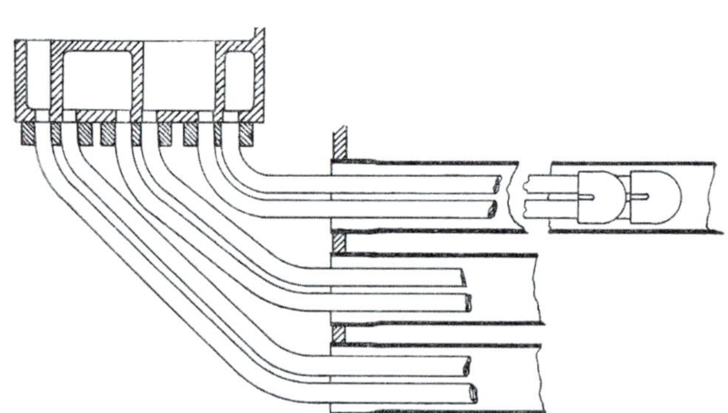

**Above: Side drawing of Schmidt superheater (header and U-tube).**

**No. 2901 *Lady Superior* was built in May 1906 with a Schmidt superheater – the first locomotive to be so fitted in the UK. This photograph was taken at Old Oak Common in July 1926, some years after a Swindon No. 3 superheater had replaced the original equipment. *F R Hebron*.**

would be insufficient to outweigh the potential savings that had been identified during tests to that date. Experiments with about nine alternative designs followed before the Swindon No. 3 type was judged the most efficient, leading to its adoption in 1908 as the company standard and to its extensive installation across the fleet. In due course, Churchward advised the Board: *'The superheaters continue to give good results and it is found that the consumption of coal is reduced by about 12%, and the consumption of water by 20%. In addition to this, the power of the engine is increased, and this is of great importance on such trains as the Plymouth "Limited" and the Fishguard "Ocean Specials", on which the full power of the present locomotive is required. The actual saving of coal is, of course, greater on the 2-8-0 mineral engines as the coal burnt per mile is much greater than that on any of the passenger services. We are therefore giving the mineral engines the preference at present in fitting the new superheaters'.*

### The Steam Cycle

A modern superheater avoids the embedded problem encountered where water and steam are in direct contact within a closed vessel. The boiling point of the former rises in relation to the pressure exerted upon it, thus:

| Pressure | Boiling Point |
|---|---|
| Atmospheric | 212° Fahrenheit |
| 85 lb/ sq in | 327° Fahrenheit |
| 225 lb/ sq in | 397° Fahrenheit |
| 250 lb/ sq in | 405° Fahrenheit |

It follows that a stronger and more expensive boiler would be required to accommodate a higher temperature at the regulator valve. However, once steam has passed this point, it is no longer in direct contact with water thus allowing its temperature to be raised further by re-directing it through the hottest part of the boiler i.e. the volume adjacent to the top and front of the firebox. This is achieved by taking steam from the regulator valve through the main steam pipe into a header mounted on the smokebox tube plate. From the header, superheater tubes set in flues take the steam back to the firebox throatplate and then by U-bend forward to the main steam pipes, and on to the steam chest. Installation of the flues and insertion of the superheater tubes therein is only possible by reduction in the number of ordinary boiler tubes and thus the related heating surface. However, this reduction in primary tube surfaces is more than off-set by the higher temperature of the steam at the admission valve. Steam processed in this fashion can be safely subjected to more heat at normal working pressures, making temperature increases viable up to between 600°F and 750°F.

Three advantages result: [1] any water that has managed to pass the regulator valve and header is immediately converted into more steam [2] condensation within the cylinder is eliminated and [3] the volume of steam increases compared with its saturated condition. The extent of this increase depends on the boiler working pressure; at 225 lb/ sq it is approximately 30%. This third factor reduces demand upon the firebox and thus fuel consumption. Passage of steam through the superheater elements helps keep temperatures in reasonable equilibrium although a hot fire when running with closed regulator (i.e. while coasting) risks their exposure to excessive heat and thus melting.

Installation of a superheater was cheaper than conversion of an existing locomotive from simple to compound propulsion (which would require a new cylinder block) or construction of a new compound locomotive with its greater mechanical complexity. Further, the fabricated nature of the equipment facilitated dimensional modification which necessitated changes only to header, throatplates and numbers/ diameters of tubes and/ or superheater flues. Later in the GWR steam story, this structural simplicity was advantageous when poorer fuel quality dictated the need for higher degree superheat with 4-cylinder 4-6-0s. Soon after conversion from saturated to superheated steam was initiated on a broad scale, it was calculated that the installation cost was recovered within 12 months through fuel savings, an impressive payback.

The moisture content in saturated steam imparts a lubricating effect upon valves and cylinders that is denied by the drier composition of superheated steam and increased rates of wear were roughly in line with temperature increases. Improved oil lubrication was thus needed which was easier to facilitate with a piston valve than with the traditional slide valve. All Churchward designs were equipped with piston valves so that in the GWR's case, the combination of slide valves and superheating was limited to 4-4-0s and Aberdares. It is uncertain whether the programme to convert these types to piston valves was ever completed.

Higher steam temperatures increased the rate of valve carbonisation so Churchward compromised with what was later regarded as a low level of superheat. With reliance primarily on good Welsh coal of superior calorific value, this policy proved sound although some observers failed to concur. Leading author and commentator Cecil J Allen remarked that by 1939, the Swindon superheater was little more than a steam dryer. However, it was generally considered that Churchward's policy yielded a sensible balance between performance, cylinder wear and valve carbonisation. During and after World War 2, deterioration in quality and consistency of coal supplies led to a revision of the GWR's policy. However, with their numbers then in decline, the surviving Saints were excluded from this exercise.

In 1914, Henry Fowler (Midland Railway) read a paper before to the Institution of Civil Engineers titled *Superheating Steam in Locomotives* which resulted in much verbal debate and correspondence from engineers at home and overseas. By then the GWR's installation programme was well under way but the tenor of the discussions implied that the No. 3 superheater could be improved upon with further analysis and investigation. In fact, other than to increase the number of elements to raise the degree of superheat, the design remained unchanged until the end of GWR steam. The pithy response evoked by these theoretically based opinions was *'... to paraphrase an Americanism, "it is the coal-pile that talks". The Great Western had been fitting superheaters for some years, but they had not done much in the way of making experiments...If engineers waited until innumerable experiments had been made to define the last degree of superheat that could be obtained, they might miss a good deal of coal in the interval."* George Jackson Churchward – the practical engineer!

**Great Western Railway No. 3 Superheater fitted within a Standard No. 1 Boiler**

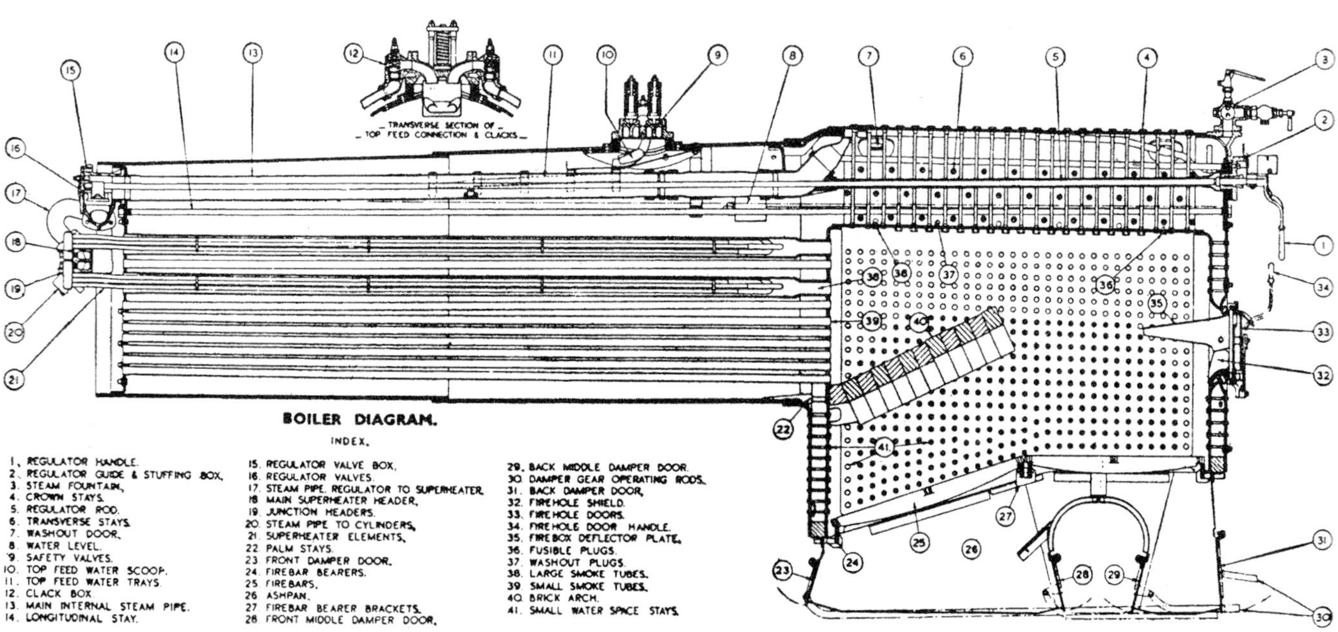

## Explanatory Drawing of the Key Components of the No. 3 Superheater

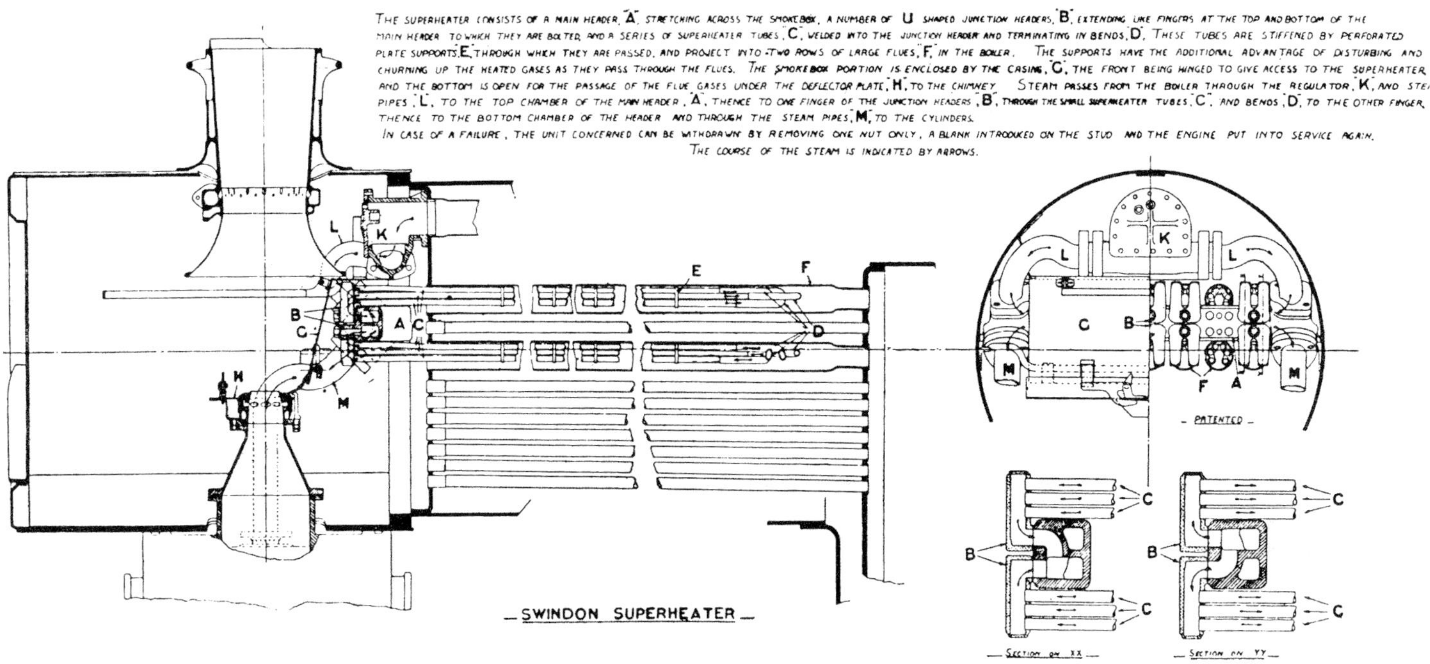


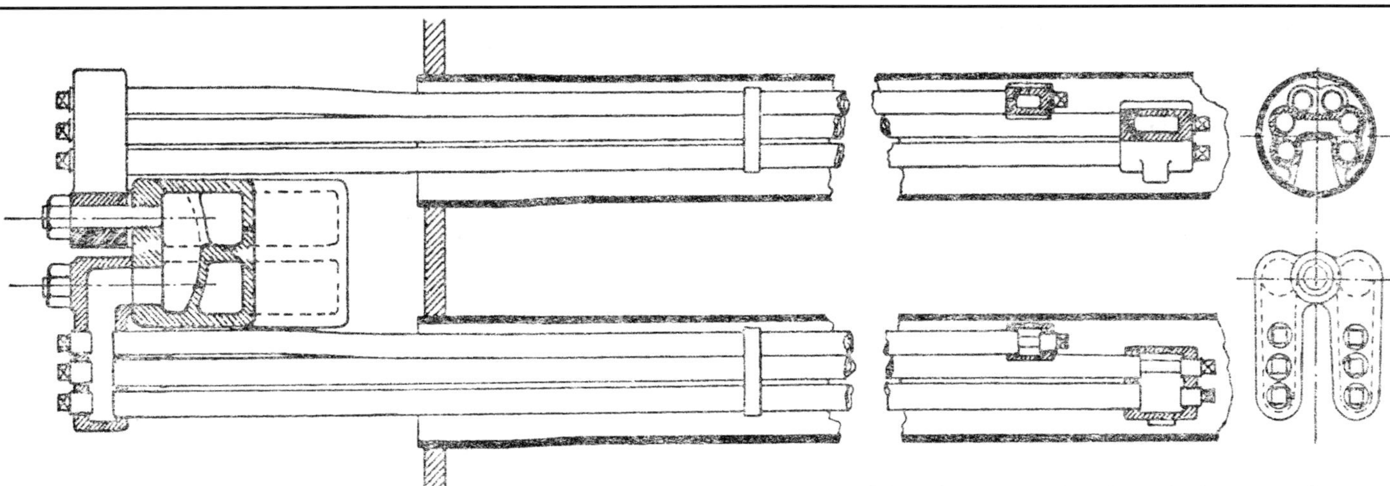

Detailed drawing of No. 3 Superheater (header and tube rear end)

### The Schmidt/ Robinson Controversy

Major attractions of the superheater are simplicity and low capital cost. The most vulnerable part of the assembly is the U-bend in the superheater tube within the flue which comprises two right-angle turns. Formation of this shape placed the metal under stress and its location in the intense heat adjacent to the firebox throatplate makes this area most prone to failure. A puncture in a superheater tube close to the firebox risks emission of high-pressure steam into the large flues, through the firebox and into the cab. Due to the U-bend's inaccessibility, ease of removal for inspection was important for maintenance. The Schmidt superheater on No. 2901 proved the efficacy of the general principle but was deficient as inspection required removal of the complete unit i.e. header and all the elements. Churchward recognised this shortcoming and his experiments focussed on means of easing flue replacement. The critical improvement was at the header whereby redesign facilitated severing of an individual element for removal and inspection. Its replacement could then be inserted and welded into place in a simpler process than was possible with the Schmidt system.

J G Robinson on the Great Central Railway reached similar conclusions leading to transition from Schmidt to the superheater type that bore his name. He personally patented his design in 1911 which was then a quite common practice among Locomotive Superintendents to augment their incomes (the highly inventive William Webb on the LNWR became wealthy through such means). However, UK patent registration of the Robinson system ignited a challenge from the Schmidt Superheating Company. A complex international dispute ensued which led to Schmidt initiating legal actions against the GCR and Robinson personally in 1914. Delayed by World War 1, the matter was eventually resolved in 1920 with the two parties merging their interests. The GWR remained immune from this controversy, presumably because although the Swindon No. 3 superheater pre-dated Robinson's patent there was no evidence of intent to pursue pecuniary reward from third parties. If this was indeed the case, then it was an interesting reflection on Churchward's open-minded character.

**Sectioned Drawing of the Smokebox and Superheater**

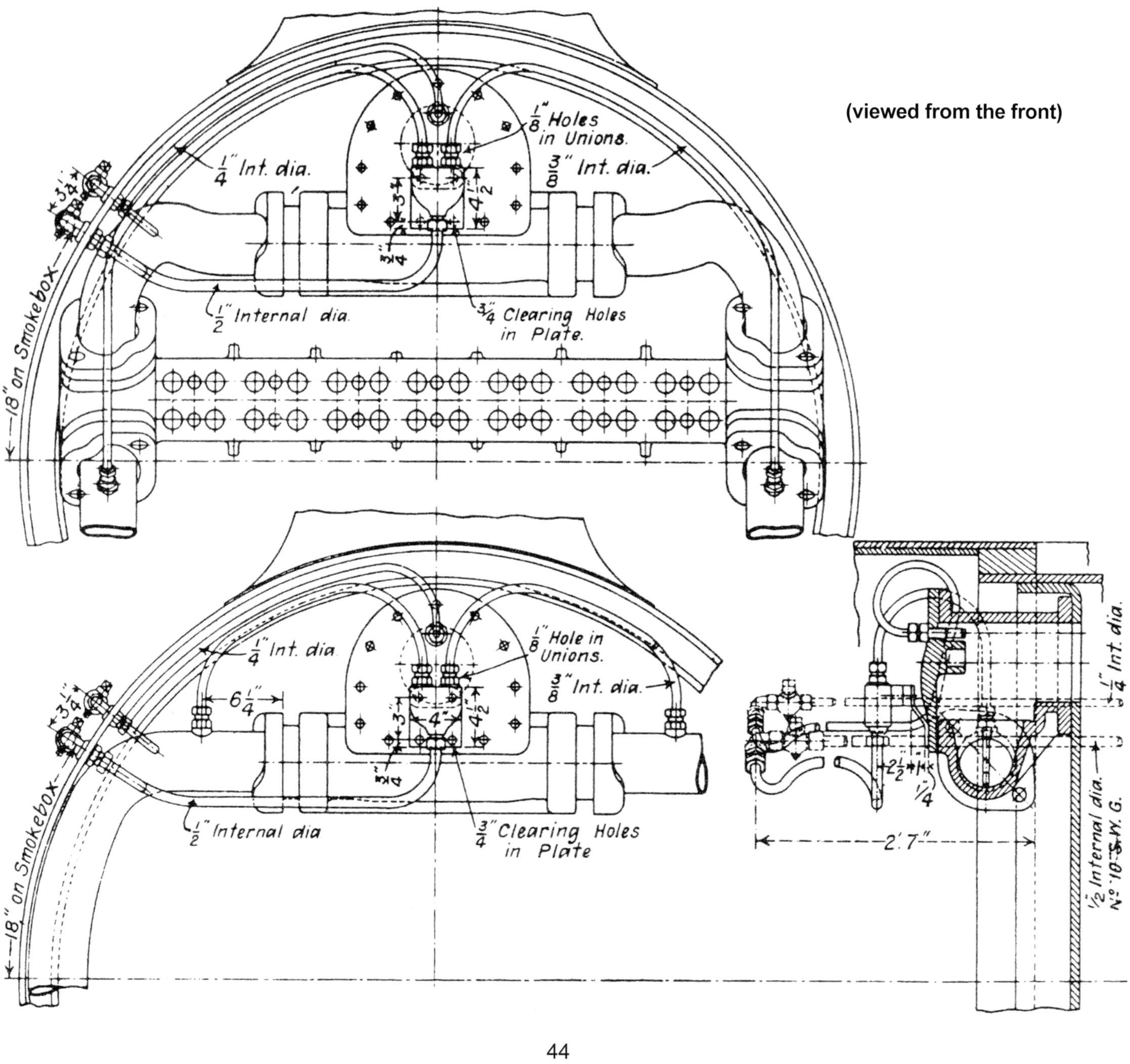

Slightly more than two years after No. 2901 had been built, installation of superheaters across the Saint Class commenced with No. 2922 in October 1908 and was completed in November 1912:

| Loco No. | Superheater Fitted | Loco No. | Superheater Fitted |
|---|---|---|---|
| **2900** | Oct 1910 | **2930** | Feb 1911 |
| **2901** | May 1906 § | **2931-32** | Oct 1911 § |
| **2902** | Feb 1910 | **2933-36** | Nov 1911 § |
| **2903** | Aug 1910 | **2937-40** | Dec 1911 § |
| **2904** | Jul 1910 | **2941-44** | May 1912 § |
| **2905** | Nov 1909 | **2945-48** | Jun 1912 § |
| **2906** | Jul 1910 | **2949-50** | May 1912 § |
| **2907** | Feb 1910 | **2951-54** | Mar 1913 § |
| **2908** | Nov 1911 | **2955** | Apr 1913 § |
| **2909** | Nov 1911 | **2971** | Oct 1910 |
| **2910** | Jan 1910 | **2972** | Mar 1912 |
| **2911** | Nov 1912 | **2973** | Oct 1910 |
| **2912** | Jun 1910 | **2974** | Oct 1910 |
| **2913** | Oct 1909 | **2975** | May 1911 |
| **2914** | Jun 1912 | **2976** | Mar 1911 |
| **2915** | May 1910 | **2977** | May 1911 |
| **2916** | Jan 1912 | **2978** | Apr 1911 |
| **2917** | Nov 1909 | **2979** | Jun 1912 |
| **2918** | May 1912 | **2980** | Nov 1912 |
| **2919** | Aug 1910 | **2981** | Oct 1910 |
| **2920** | Oct 1909 | **2982** | Nov 1912 |
| **2921** | Jan 1912 | **2983** | Oct 1910 |
| **2922** | Oct 1908 | **2984** | Jun 1912 |
| **2923** | Apr 1911 | **2985** | Apr 1910 |
| **2924** | Jun 1910 | **2986** | Feb 1912 |
| **2925** | Oct 1909 | **2987** | Apr 1912 |
| **2926** | Nov 1910 | **2988** | Aug 1910 |
| **2927** | Oct 1910 | **2989** | Jul 1911 |
| **2928** | Feb 1911 | **2990** | Mar 1911 |
| **2929** | Aug 1912 | **2998** | Sep 1911 |

§ Installed when built.

By December 1912, about 750 superheater units had been manufactured at a rate of about four per week at Swindon and they were mainly fitted to new or rebuilt Standard Nos. 1, 2 & 4 boilers. The speed and scale of this programme was remarkable and assuming that the estimated *circa* one-year payback was achieved, the aggregate financial benefit must have been considerable. Comparison with the contemporary superheater programmes of other companies demonstrates just how advanced was the pre-Grouping GWR in this regard:

| **Railway Co.** - fitted by year end: | **1910** | **1914** | **1921** |
|---|---|---|---|
| **Great Western** | 29 | 801 | 1472 |
| **Caledonian** | 5 | 33 | 103 |
| **Great Central** | 1 | 142 | 373 |
| **Great Eastern** | - | 37 | 178 |
| **Lancashire & Yorkshire** | 27 | 87 | 249 |
| **London & North Western** | - | 158 | 613 |
| **London & South Western** | - | n/a | 86 |
| **London Brighton & South Coast** | 15 | n/a | 50 |
| **Midland** | - | 62 | 257 |
| **Total** | **77** | **1320** | **3381** |
| **GWR (Total %)** | **38** | **61** | **44** |

As at 31 December 1921, it was estimated that 47% of the GWR's fleet had been superheated. However later trends suggested that the campaign had been over-enthusiastic as the cost-benefit equation was less impressive for motive power engaged in duties typified by modest loads, low speeds, lengthy periods stationary, or short journey distances. For example, 318 0-6-0Ts built up to March 1901 carried superheaters at some stage in their careers. With resumption of fleet renewal of 0-6-0PTs by construction and purchase in 1929, all but ten (Class 94xx Nos. 9400-9) were saturated.

## Top Feed

Prior to introduction of the domeless tapered boiler, various locations were used for the clackboxes by which feed water was admitted to the barrel. During the Dean era, the most obvious position was externally on the barrel side toward the smokebox end, and formed of brass, usually highly polished in the fashion of those times. Alternative positions were either on the firebox back or sides. With the early Saints, two clacks were fitted side-by-side behind the smokebox, attached to two branches of a 'Y' casting bolted to the bottom of the barrel through which feedwater entered the boiler. A short hood was attached the stem of the Y fashioned to direct incoming water horizontally toward the firebox. The cool water thus entered clear of the tubes and its direction assisted circulation. However, the temperature difference strained the boiler plates plus any oxygen dissolved in the water that had been captured from the atmosphere at the injector stage contributed to corrosion and attendant pitting of the plates. Observation of individual cases revealed that damage was worse with boilers that had been supported mid-length by a frame stay so mechanical stress added to the deterioration.

Experimental attempts to devise a means of feeding water from above were conducted in Germany in 1863, followed by similar efforts in France in 1890. Neither of these initiatives made much practical progress so beyond general recognition of the potential benefits, it seems that top feed as introduced by Churchward in 1911 had no specific precedent. Feedwater from the injectors was piped to immediately below the safety valve and then by right-angled turn vertically upward and around the barrel to clack boxes installed either side of the bonnet, and enclosed within the familiar 'ears'. By sharing the opening in the boiler that accommodated the safety valve mechanism, a further penetration of the barrel was eliminated. Feedwater fell onto a sloping perforated tray inclined toward the firebox which ensured its distribution as spray at a temperature estimated at about 40°F below that of the boiler water. Pitting was thereafter confined to the tray which was easily replaced as required. Holcroft recalled that 'bench testing' was conducted in the paved yard adjacent to the Swindon drawing office whereby measured water volumes were directed by hose onto a test tray which was progressively re-fashioned to determine the optimum shape for water distribution.

Through this simple process, the efficiency of water supply was improved with substantial cost savings. The system was installed on all subsequent tapered boilers, plus some of the parallel variety (e.g. Dean Goods and from the 1940s, 0-6-0PTs). This may have been another example of Churchward's expansive recognition of the potential residing in foreign engineering practices.

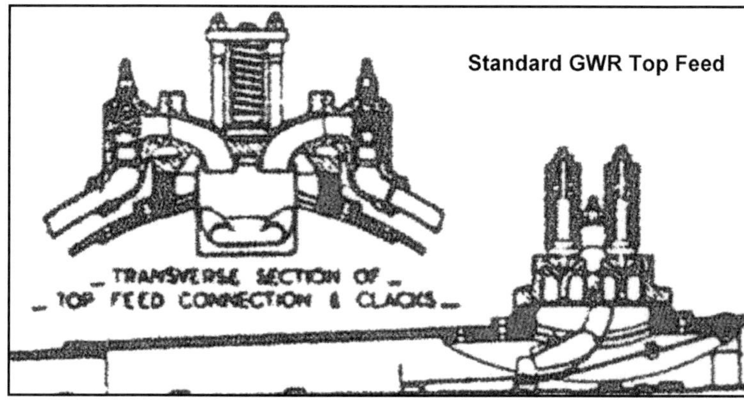


**Standard GWR Top Feed**

**Above: No. 2980 *Cœur de Lion* had started life as a 4-4-2 in May 1905 with half cone boiler, vertical front fall and straight rear running plate. It acquired a full cone superheated boiler in November 1912 and was rebuilt as a 4-6-0 in January 1913, apparently during the same works visit. With the change of wheel arrangement, it acquired curved falls and valences front and rear in the style inaugurated by No. 2911 *Saint Agatha* when built in August 1907. It is recorded taking water at Shrewsbury probably on a West to North train in the mid 1920s, with steam escaping from the tall safety valve bonnet. Note also the Automatic Train Control (ATC) shoe has been fitted below the front buffer beam.** ***W H Whitworth (courtesy Stephenson Locomotive Society).***

# REBUILDING PROJECTS

During the inter-war years, the Saints were engaged in three rebuilding projects which shed some light on the intricacies of Swindon's policies during a period when in contrast to the enterprise of the Churchward years, embedded conservatism engendered stagnation in motive power affairs. Post-war, a small but crucially significant cadre of ex-GWR locomotives considered worthy of preservation was assembled. During the Saints' final operational years, failure to include a class member within that select group was deeply regretted within the enthusiast community. The fourth project, initiated long after the GWR ceased to exist, rectified that omission.

### The Super-Saint

Churchward's last design was mixed traffic 2-8-0 No. 4700, built with a Standard No. 1 boiler and 5' 8" driving wheels. Surprisingly, this boiler proved inadequate for the new engine's intended duties which was something of an embarrassment. The *Great Western Railway Magazine* for October 1921 reported that *'… the locomotive was given an exhaustive trial and it did good work on semi-fast passenger services among which were the following trains …* (NB There followed a list of services although significantly none included heavy overnight fast freight trains for which the design was primarily intended) *…At the time the engine was put into service the proper boiler was not ready, and it was sent out fitted temporarily with a No. 1 standard boiler. A short time ago it was withdrawn from traffic and sent into the Swindon factory to receive her proper boiler, which is larger than that first placed upon the frames…'.*

In January 1922, the same magazine reiterated that the standard No. 1 boiler had been temporary and that *'In July last a No. 7 standard boiler was substituted. The steam admission pipes to the cylinders are now taken through the smokebox and coupled externally to the cylinders.'* There was a disingenuous flavour to these reports. For example, how could 'good work' morph into the need for a new boiler design after two years' service? Further, why with a new design should it have taken so long for the allegedly temporary No. 1 boiler to be replaced?

As it turned out, the No. 7 boiler was so effective with Class 47xx that diagrams were prepared for it to be fitted to Stars, Saints and Class 28xx. These proposals were unfortunately stymied beyond the drawing board

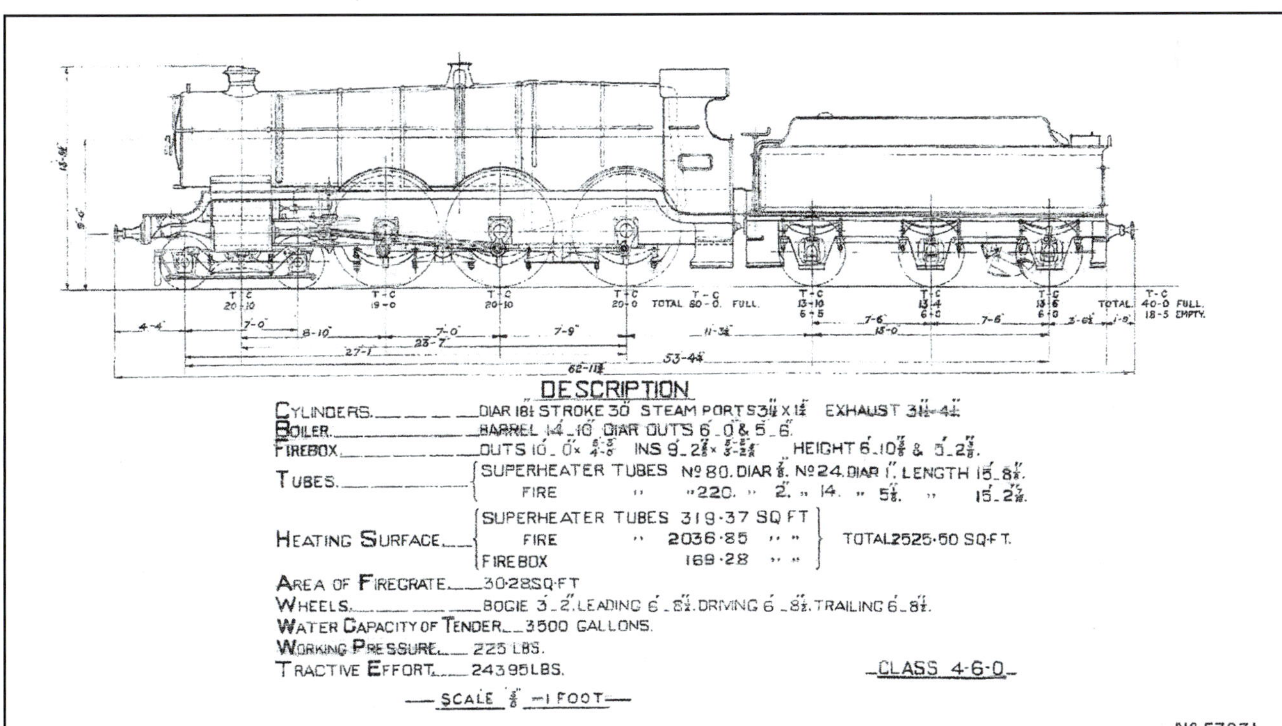


**Above: Diagram for proposed Saint fitted with Standard No. 7 boiler. Despite the *GWR Magazine* referring to outside steam admission pipes with the boiler installed on 2-8-0 No 4700, there is no indication of this feature on this drawing (nor on those prepared concurrently for the Star and 28xx classes).** ***GWR Drawing Office Official.***

stage due to constraints imposed by the Civil Engineer concerning weight and loading gauge. The issue of bridge loadings was discussed in *Western Times* Issue 3 in connection with the planned introduction of the Kings and the revelation that comparatively little civil engineering work was needed to enable their use Paddington-Plymouth. *Western Times* Issue 12 alluded to the extraordinary boiler performance of *The Great Bear* on a working to Bath. It can only be speculated what opportunities were lost simply because of poor communication between the CME and his civil engineering counterpart. Nevertheless, of the three diagrams using the No. 7 boiler, that in respect of the Super-Saint despite its formidable potential was probably the least viable on account of the hammer blow issue.

### No. 2925 *Saint Martin*: The First 'Hall'

Churchward's ground-breaking Standard Locomotive scheme of 1901 envisaged only four driving wheel diameters (6' 8½", 5' 8", 4' 7½" and 4' 1½"). The largest was first used in those 4-4-0 classes intended for express passenger duties, followed by the Saints and finally the 4-cylinder 4-6-0 family. With the policy decision to favour 4-cylinder 4-6-0s for high speed passenger duties, the focus switched to 2-cylinder propulsion for mixed traffic work.

Churchward had anticipated the need for a 4-6-0 with 5' 8" driving wheels for slower and more varied services. In the immediate term these duties were covered mainly by 5' 8" 4-4-0s and then by the versatile 2-6-0 Class 43xx, built in large numbers from 1911 as a late addition to Churchward's scheme. By the mid-1920s, operating demands were stretching the capacity of the moguls' Standard No. 4 boiler, and due to their poorer adhesive qualities, also the 4-4-0 fleet. Therefore in 1924, No. 2925 *Saint Martin* was selected for experimental conversion to mixed traffic status by fitting 6' 0" diameter driving wheels, despite the Running Department's advocacy for a 4-6-0 with 5' 8" driving wheels and No. 1 boiler, exactly as Churchward had planned. However, Collett's caution concerning fresh designs led to a minimalist response.

The Saint's mainframe and boiler were retained while 6' diameter driving wheels were installed with commensurate adjustment in the cylinder alignment, splashers etc, resulting in a 4¼" lowering in the boiler pitch to 8' 1¾". The other major change was installation of a side window cab of the style inaugurated with the Castle Class. Collett took his time before multiplication of the type with the first eighty production Halls delivered under Lot 254 between 1928 and 1930. With these engines and up to No. 6958 (Lot 340 of 1941-3), the boiler pitch was restored to 8' 6". No. 2925 was renumbered 4900 on 7 December 1928 and withdrawn in April 1959, 5½ years after the last pure Saint was taken out of service.

Use of a fresh driving wheel diameter was a key break with the Churchward format. The 5' 8" dimension had proven eminently suitable for mixed traffic tender locomotives (classes 43xx and 47xx) but use of this size would have required more extensive structural modifications which Collett chose to avoid. Re-wheeled No. 2925 proved a success but when key parts recovered from withdrawn members of Class 43xx were adapted for the 5' 8" 4-6-0 Grange Class in 1936, Churchward's original vision was belatedly fulfilled. The operational reputation of the Grange class suggested that it was an improvement upon the Halls.

Above: Looking in a reasonably presentable external condition (apart from the somebox door scorching), No. 4900 *Saint Martin* awaits its next turn of duty at Old Oak Common on 23 September 1956. Throughout its time as a Hall the locomotive was paired with both 3,500 and 4,000 gallon tenders, before withdrawal on 3 April 1959.
*R C Riley (RCRC 4219).*

Right: No. 4900 *Saint Martin* at Landore shed on 9 September 1951. Despite the angle, retention of the lower boiler pitch is clear, as is the shorter outside steam admission pipe. Note also the tell-tale absence of brass beading on the splashers, another indication of its heritage. As can be seen from the 83A shedplate, the locomotive was allocated to Newton Abbot for a short period between April and October 1951, its earlier home as a Saint.
*R C Riley (RCR 3436).*

Opposite Page: Hall No. 2935 *Saint Martin* as originally converted from a Saint class locomotive, fitted with the new side window cab and 6' 0" driving wheels but retaining its inside steam pipe-type boiler. It is recorded here on 10 August 1925 at Plymouth North Road piloting the second part of the 'Up Limited' with a Castle class. From 1929 onward, there was extensive debate among observers over when and whether this locomotive acquired outside steam pipes. Without seeking to add to the confusion, it seems that the locomotive might have swapped identities with No. 4901 *Adderley Hall* for photographic or promotional purposes, or even for charter work connected with the 'St. Martin Preserving Co. Ltd'. The key determinant should have been the boiler pitch but this was before the Halls assumed prominence in the fleet. This has resulted in uncertainty regarding the exact dates on which the locomotive acquired outside steam pipes, and when the number was changed to 4900. *F H C Casbourn (courtesy Stephenson Locomotive Society).*

### No. 2935 *Caynham Court*: Rotary Valves

Collett appeared to favour the Saints for fresh initiatives over any 4-6-0 design directly attributable to him. His motives were unclear but in view of his secretive nature, he might have been unwilling to present any of his own designs for critical comparative analysis. An example of this attitude was his surprise unveiling of the up-graded Stationary Test Plant which involved the use of No. 2931 (see page 54).

Another aspect of his personality was deep-rooted antipathy towards the commercial locomotive manufacturing sector. He was convinced that the Big Four's workshops produced locomotives inherently superior to the typical motive power delivered by private industry. This belief ignored the reality that the majority of commercially-sourced engines were export sales that had to compete against foreign competition and to operate under conditions often vastly more arduous than anything encountered on the GWR system. Further, his subordinates were forbidden from joining the Institution of Locomotive Engineers (which he dismissed as a forum for 'commercial travellers') despite the opportunities for cross-fertilisation of fresh ideas.

In this climate, the rebuilding of No. 2935 *Caynham Court* with Lentz Rotary Cam poppet valves and new cylinders in May 1931 appeared at odds with Collett's technological isolation but recent research has revealed something of the circumstances that led to this atypical exercise. In 1886, the general engineering firm of William Beardmore Ltd was created out of a forging business that had been established at Parkhead, Glasgow in 1835. Under William Beardmore's dynamic leadership the company's activities expanded both rapidly and diversely. By the close of World War 1, it was a leading shipbuilder which also produced armour plate, guns, tanks, airships, and motor vehicles. With trade declining, in 1920 the company diversified into locomotive construction which included acquisition of the UK rights to manufacture and distribute Caprotti valve gear under the 'Beardmore-Caprotti' label. By the early 1930s with the group on the verge of bankruptcy, its various activities were closed down or sold off.

In 1930 the company converted 4-cylinder 4-6-0 Nos. 401/ 6 of the Great Southern Railways, Ireland to 2-cylinder form with Beardmore-Caprotti valve gear. Although never formally confirmed, it is believed that most, possibly all, of the cost was borne by Beardmore in the hope of attracting further revenue-earning contracts. The GSR's condition for most of its 20-year existence was financially weak and while enjoying advantageous terms with this transaction, any expectation of lucrative consequential business for Beardmore proved a matter of hope over reality.

**Above and Opposite Page Top: The official front three-quarter and side profile views of *Caynham Court* on 1 June 1931, following its rebuilding at Swindon Works during the previous month with new cylinders and rotary poppet valve gear. The selection of this locomotive for the modification seems to have been purely down to it being next in line for an overhaul.** ***GWR Official.***

That Beardmore attempted a similar promotion with the GWR is confirmed by the following minutes of the Locomotive Committee:-

**4 October 1928** – *'Under circumstances represented the Committee agreed to facilities being extended to William Beardmore & Co Ltd for trials to be made on one of the company's 2900 'Saint' class locomotives of a boiler built by the firm for use with pulverised fuel and also a set of cylinders with Caprotti valve gear on another similar locomotive, the cost of the experiments being borne by Messrs Beardmore. The arrangement will be such that the boiler could be transferred later, if desired, to the engine fitted with the Caprotti valve gear'.*

**29 May 1930** – *'Referring to minute No. 9 dated 4 October 1928, it was reported by the CME that William Beardmore & Co Ltd were unwilling to carry out at their own expense the experiments with Caprotti valve gear and that matter had been allowed to drop. A poppet valve for locomotive use had been developed by Associated Locomotive Equipment Ltd and the committee approved the purchase from them for experimental purposes of one set of valves with operating gear and internal drive for £860 f.o.r. (sic) Swindon'.*

The sequence of events suggests that Beardmore's approaches to the GSR and GWR were more or less concurrent and that the interval between promotion and rebuild of the GSR engines was taken up in equipment design and manufacture. The GWR was adept at taking advantage of subsidies in any form as witness the 0-6-0PT Class 57xx construction programme and the financial assistance for civil engineering works provided by the Development (Loan Guarantees and Grants) Act 1929. The opportunity of 'gaining something for nothing' through the Beardmore offer probably overcame Collett's antipathy toward incursions by outside influences.

So far as Beardmore was concerned, the change of heart vis-à-vis the GWR probably reflected realisation that the sub-standard GSR 4-cylinder Class 400 offered greater scope for improvement and hence promotional advantage whereas the chances of improving a Saint were distinctly slimmer. The parallel offer to supply a boiler designed to burn pulverised fuel was strange, possibly motivated by the exigencies of the 1926 Coal Strike. After World War 1, the Great Central had conducted tests with 2-8-0s modified to burn differing forms of pulverised and colloidal fuel with varied results, none which supported wider adoption of this means of combustion.

The only tangible outcome to these deliberations was acquisition of the Lentz equipment and its installation in No. 2935. A key weakness with piston or slide valves activated by Walschaerts or Stephenson Link Motion is that the ports fulfil the dual role of admitting/exhausting steam to/from the driving cylinder. Because of its expansive properties, the volume exhausted is greater which exerts an inevitable throttling effect upon the flow. To counter this impediment, the Caprotti and Lentz systems employ oscillating or rotary valves where those for exhaust have the greater surface area. Several other advantages over the traditional piston valve claimed:-

**Above: The cylinder casting with rotary camshaft layout for No. 2935.** ***The Railway Gazette.***

**1)** Overall total weight reduction of the complete assembly compared with piston valves.
**2)** Significantly lessened potential for wire-drawing (steam leakage past the valve head).
**3)** Fewer non-productive movements of mechanical parts.
**4)** Wider possible range of cut-off adjustments.
**5)** Reduced exposure to violent temperature changes that might induce metal fatigue.
**6)** Cam profiling allows sharpened valve opening followed by gentle closing thus reducing the impact.

Installation of the Lentz system in mid-career was costly with the concomitant pressure upon satisfactory amortisation but promoters claimed that a 20% reduction in fuel consumption was possible and certainly in some countries, impressive savings were achieved. The system was more reliable than conventional valve gears but when adjustments or repairs were necessary, specialist skills were required.

The UK rights were originally held by Davey Paxman Ltd but these had been acquired some years earlier by Associated Locomotive Equipment Co Ltd who supplied the equipment, as described in *The Railway Gazette* for 12 June 1931. Standard ALE valves were used but with significant ancillary modifications. Instead of the normal outside drive, a gearbox was mounted between the frames on the driving axle from which a shaft extended forward to engage with an enclosed transverse shaft that ran in double-purpose ball-bearings. At either end of the transverse shaft, gears drove short shafts on either side of the engine forward to the cam boxes. Thus the drive was delivered through a series of bevel gears between the driving axle and the rotating camshafts. Universal joints were inserted in the main drive shaft from the axle and in the short cam shafts. An obvious cost factor with this arrangement lay in the number of bevel gears that had to be specialist-machined as such items were lay outside the scope of normal railway engineering work.

Although standard ALE valves were used, their position was reversed from the normal arrangement as the exhaust valves were placed outside and the admission valves inside. This allowed the adoption of specially enlarged exhaust ports and passages; inlet port diameters were 6 3/16" and exhaust were 8". In forward gear the cam mechanism permitted nine different rates of steam admission through cut-offs ranging from 10% to 85%. To alter the cut-off or to engage reverse gear, the camshafts were moved transversely by a system of levers that connected with ball-thrust boxes fitted to the inner ends of the cam shafts. The outside extremities of these levers engaged with a rack and pinion actuated by a worm and wheel gear, in turn connected to a rotary

**Above: Detaill of the Lentz valve gear and revised cam covers for the Corliss valves. Photograph dated July 1947. *R C Riley (RCR 1036).***

**No. 2935 *Caynham Court* stands on the coaling road at Swindon shed where it was allocated post rebuilding until withdrawal in December 1948. The removal of the lamp bracket from the top of the boiler to the smokebox door suggests a date between 1935-1939.** ***W H Whitworth (courtesy Stephenson Locomotive Society).***

shaft extending from the front of the engine on the right side of the boiler and connected to a hand-operated reversing wheel in the cab that was similar to the normal type. Despite the promoters' claims, the conversion added to the total locomotive weight, probably due to the conversion of an existing locomotive. Claimed weight savings might have been realised had the modifications been implemented during construction (to a new Hall 4-6-0 perhaps?). Weight comparisons [tons]:

| Loco Nos. | 2998 | 2971-90 | 2902-10 | 2935 Modified |
|---|---|---|---|---|
| Bogie | 16.4 | 16.6 | 17.2 | 19.4 |
| 1st Driving Axle | 16.4 | 18 | 18 | 18.3 |
| 2nd " " | 18 | 18.6 | 18.4 | 18.6 |
| 3rd " " | 17.5 | 18.5 | 18.4 | 18.7 |
| Total | 68.3 | 71.7 | 72 | 75 |

Road trials were conducted over about 18 months from May 1931. Little was published about the results but findings were summarised in a paper read by Samuel Ell in 1933 which concluded that the hoped for improvements were unrealised. Further, locomotive crews opined that the engine was about 1-2 coaches weaker than a normal Saint. During the tests, steam leakage was encountered and several modifications were tried including Corliss valves (a patented design of American origin) whose presence was evident in the different shape of the cam box housings above the running plate. Collett appeared to lose interest in the project fairly quickly so after the Corliss installation, the engine ran in this condition until withdrawal.

Collett had been intent on remaining in post until aged 70 years which he almost achieved and R A G Hannington, Swindon Works Manager, was widely regarded as his anointed successor. However, Hannington died in a freak accident on 26 June 1937 and with his untimely passing, F W Hawksworth assumed that mantle. As was later proved, 'Big Freddie' was receptive concerning ideas outside the Swindon tradition and there is fragmentary evidence of discreet contact between him and ALE in the late 1930s. The discussion apparently centred on the fitting of higher-lift cams to improve the steam flow. This would have been a simple low cost modification, which never eventuated due to Collett's reticence. Nevertheless, this suggests residual interest in the Lentz system although no trace has been found of concrete attempts regarding its further application.

However, there was an intriguing postscript to the *Caynham Court* story. In early BR days, a survey was conducted of revolutionary initiatives that had been sponsored by the Big Four to evaluate whether any deserved further investment. This exercise was carried out in sympathetic fashion as exemplified by the latitude extended toward proving the viability or otherwise of the Bulleid Leader Class. No. 2935 fell within the survey's remit and investigation concluded that the cross-sectional area of the admission ports at 19 square inches was unduly small relative to the driving cylinder dimensions. With steam flow restricted at the start of the piston cycle, new design of cams would have been unlikely to offset the key issue of port size let alone address any other shortcomings. Redesign of the ports would have been unduly costly for a 37-year old member of a class then approaching the final phase of its withdrawal programme.

**No. 2931 *Arlington Court* sat on the Swindon Works Stationary Test Plant. This locomotive appears to have been a regular choice for testing purposes, spending much of 1935 being utilised in addition to its presence during the CME visit discussed below.**

### No. 2931 *Arlington Court*: Stationary Test Plant, Swindon

Stationary testing plants were developed in the USA from 1891 and shortly after completion of the Pennsylvania Railroad facility at Altoona in 1904, Churchward proceeded with a similar albeit smaller version at Swindon. The 'home trainer' was intended for testing motive power capacity under controlled conditions and for running-in locomotives following works attention. The facility was constrained by its inability to absorb the power output of the GWR's largest locomotives, a curious inconsistency considering his objective to produce locomotives capable of generating 2,000 drawbar horsepower at 70 mph. In practice the facility in its original form was little used.

Recognising the need for a fully-fledged testing plant in the UK, following extended discussions the LMS and LNER formed a committee in 1937 to explore creation of a facility as a joint venture under the auspices of the Association of Railway Locomotive Engineers. By then, Collett had little involvement in the affairs of ARLE but apparently remained informed about what was afoot. Knowing that a facility was being planned (as ultimately came to fruition at Rugby), he uncharacteristically invited his fellow CMEs to Swindon for a tour of the works which included presentation of No. 2931 going nowhere at high speed on the fully refurbished test plant. In this case, rebuilding concerned the facility and the Saint was merely the instrument for its demonstration. Apparently the rebuilding had been in hand for some time, at considerable expense and in total secrecy. Collett's joke apparently had the hoped-for impact upon his audience but the plant still saw little use until BR days. The choice of No. 2931 was intriguing and perhaps reflective of Collett's opacity in design matters. His visitors would have naturally probed him for technical details for comparative purposes. Employment of a representative of an ageing class then subject to withdrawal would deflect enquiry away from just how well a modern GWR 4-6-0 could perform.

### No. 4942 *Maindy Hall*

Regret over the loss of No. 2920 *Saint David* and the notional simplicity of reversing the Saint to Hall transition motivated acquisition of No. 4942 from the Barry scrapyard. The original objective had been replication of a late condition Saint. However, progress in the preservation movement advanced the feasibility of a more extensive rebuild that resulted in creation of No. 2999 *Lady of Legend* as an early version 4-6-0, with the facility for retro-conversion to Atlantic form. Halls, Modified Halls and Manors are well represented in preservation and 'new build' enterprise has since seen No. 6880 *Betton Grange* added to this distinguished group. With work well advanced on a 'new' No. 1014 *County of Glamorgan*, future generations before long should be able to appreciate examples of the full range of GWR express and mixed traffic 2-cylinder 4-6-0s.

# TENDERS

The type of tender generally associated with the Saints is the ubiquitous Churchward 3500-gallon design, which all of the class were assigned at some point. The story is as usual not so straightforward, and there follows an examination of the various types known to have been coupled to the class. This information is extracted from a combination of the official Engine Record Cards and extensive photographic study.

Editors Note: The complex subject of Great Western Railway tenders is to form the basis of a future *Western Times Special* issue.

### Dean 4000-gallon

This tender type was designed during William Dean's tenure, with construction continuing under Churchward. In total only twenty examples were built:

| Lot No. | Tender Nos. | Tender Built | Loco No. | Loco Built |
|---|---|---|---|---|
| A46 | 1456-61 | Apr to Aug-00 | - | - |
| A51 | 1509-18 | Sep to Dec-01 | - | - |
| A53 | 1539 | Jun-02 | **100** | Feb-02 |
| A55 | 1560 | Apr-03 | **98** | Mar-03 |
| A56 | 1561 | Jul-03 | 97 | Jun-03 |
| A60 | 1582 | Feb-04 | **171** | Dec-03 |

Tender Lots A53/ 55/ 56/ 60 were specifically issued for the first three 4-6-0s and the prototype 2-8-0 as listed in the table opposite. No explanation has been found for the disparities in dates. Early photographs of No. 100 show this engine uniquely attached to a tender with coal rails (page 11). However, by the time it was photographed at Paddington bearing the standard form of *William Dean* nameplate (page 26), its tender was of the side fender type. The remaining three tenders in this category were fitted with side fenders from new.

Tender No. 1560 originally coupled to Locomotive No. 98 was later modernised with new frames and the Collett-style continuous wraparound fender. It was recorded coupled to locomotive No. 2938 at Swindon in 1952. In this form it closely resembled a Collett 3500-gallon tender (Lot A118 Nos. 2242-68 built April 1929 to March 1930).

Several of the first 'Scott' production series of the class (Nos. 172-90) are known to have ran with Lot A46 and A51 tenders. These vehicles had a 7-ton coal capacity, weighed 19 tons empty and 43 tons full, with a 7' 6" x 7' 6" wheelbase. Some also found use behind later batch locomotives, such as Lot. A51 tender No. 1513 which ran with No. 2931 *Arlington Court* between September 1921 and March 1923. Upon withdrawal in October 1953, the records also show that No. 2920 *Saint David* was paired with A51 No. 1509.

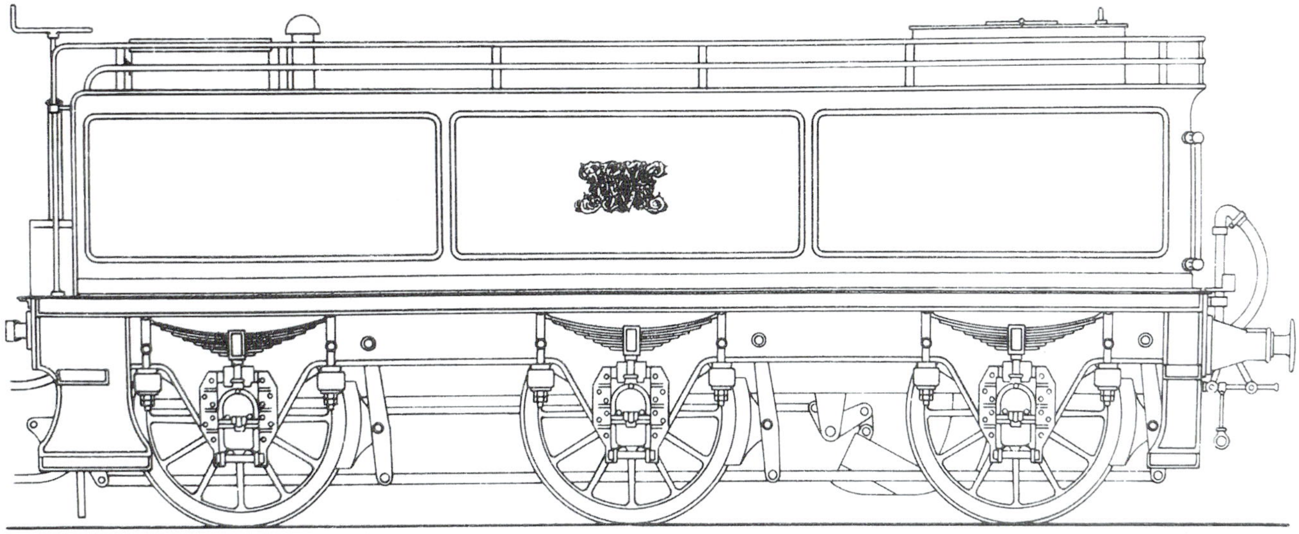

**Above: Drawing of the William Dean designed 4000-gallon tender, as originally constructed with coal rails (Lots. A46/ 51/ 53). In this form the tender is only known to have been paired with No. 100 and photographic evidence suggests this was only for a short period of time, before it was replaced with an example modified with the side fender. The design continued to be built following Churchward becoming CME at Swindon (Lot. A55/ 56/ 60), but these and subsequent modifications to the earlier builds included the characteristic fender.**

**Above: No. 178 *Kirkland* is recorded at Plymouth Laira shed *circa* 1906, paired with a Dean 4000g tender incorporating the side fenders. The tender displays the early 1900s standard livery of middle chrome green embelished with full lining to the main tank side (in three panels), fender and even on the rear face.** ***A R Kingdom (courtesy Stephenson Locomotive Society).***

## Churchward 3500-gallon

This type totalled 649 examples built between 1905-30 and was the design most commonly associated with the class. Early tenders were identified by short side fenders, but after 1910 longer side fenders were installed as will be apparent from the images.

New locomotives from Nos. 2901-55 were attached in sequence to new tenders as follows:

| Lot No. | Tender Nos. | Tenders Built | Fender | Loco Nos. |
|---|---|---|---|---|
| A67 | 1644-53 | 1906 | Short | 2901-10 |
| A75 | 1735-54 | 1907 | Short | 2911-30 |
| A83 | 1831-40 | 1911 | Long | 2931-40 |
| A86 | 1871-89 | 1912 | Long | 2941-50 |
| A89 | 1911-15 | 1913 | Long | 2951-55 |

**Churchward 3500g tender (post 1910 build with longer side fender)**

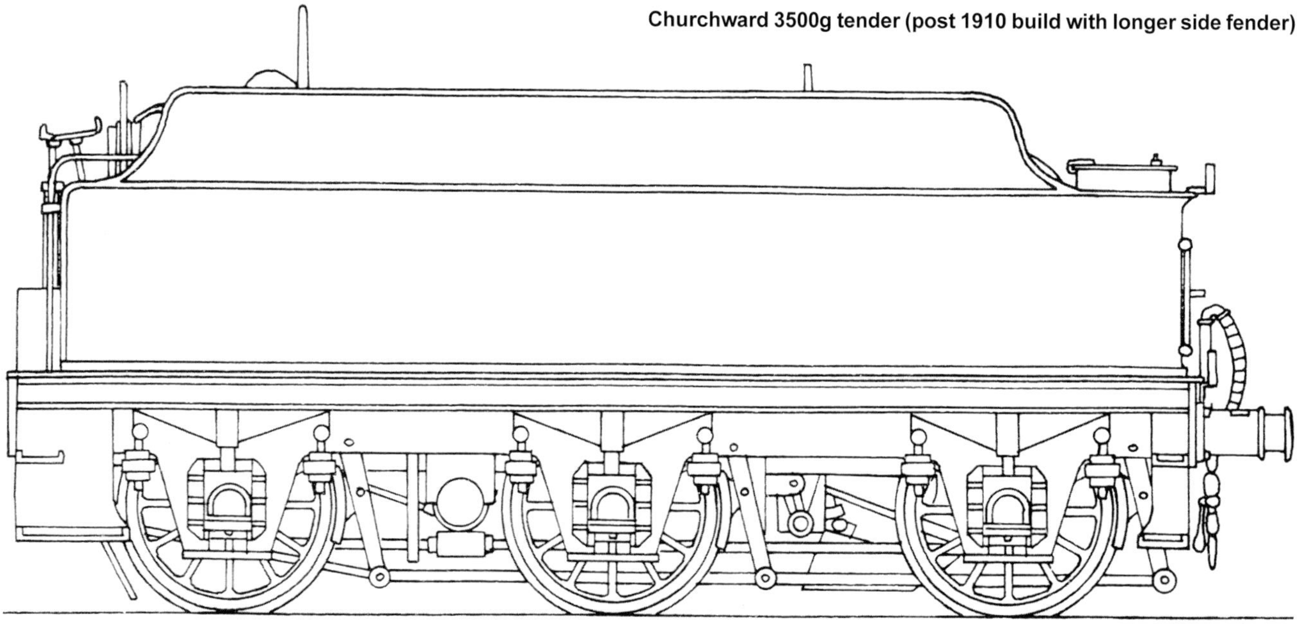

Left: The short fendered Churchward 3500-gallon tender is recorded paired with No.174 *Lord Barrymore.* This early example features flush riveting on the tank sides, a laborious process involving grinding during construction to produce a more aesthetic appearance. From around 1923 onwards, protruding snaphead rivets became the accepted rule, possibly as a result of a Collett economy measure.

The tender side fenders which became a standard feature under Churchward, reduced coal loss while in motion. The shorter version fitted until 1910 were some 15-feet in length, which were subsequently increased to 18-feet as seen in the images below. *Photomatic.*

Above: The classic Churchward pairing of No. 2915 *Saint Bartholomew* and a long fendered 3500g flush riveted tender. In this ex-works c1920 condition the engine has gained a cast iron chimney and lost the splasher beading, but retained its bogie brakes. *GWR Official.*

Below: A snaphead riveted Churchward 3500g tender coupled to No. 2920 *Saint David* at Swindon on 15 June 1952. *R C Riley (RCR 3675).*

### Churchward 3500-gallon eight-wheeled No. 1755

This unique 3500-gallon tender had been built to Lot A76 in February 1908 for No. 111 *The Great Bear*, but with the Pacific's withdrawal for rebuild in 1924, it was subsequently noted paired with 4-4-0 Counties, Saints and Stars. It is known to have worked with No. 2914 (Oct 31-Jan 33), No. 2916 (Jan 33-Mar 35) and No. 2912 (Nov 35-Jun 36). The tender was withdrawn in July 1936 and was cut up at Swindon by October of the same year.

**Above: Churchward eight-wheeled tender No. 1755 seen attached to No. 2912 *Saint Ambrose* at Plymouth Laira shed in early 1936.**

### Churchward / Collett 3500-gallon 'Intermediate'

The final ten 3500-gallon Churchward tenders built under Lot A112 (Nos 2374-83) between July 1925 and January 1926 differed from the remainder in having higher sides and almost full length side fenders. They appeared larger than the standard form although their nominal carrying capacity remained unchanged.

**Above: Churchward 3500-gallon tenders were built in considerable numbers between June 1905 and January 1926 but the final batch of ten built under Collett had higher sides (Lot A112). An example of this batch is seen here coupled to No. 2971 *Albion* at Swindon shed on 6 June 1937. At least ten class members ran with this design type. *W A Camwell (courtesy Stephenson Locomotive Society).***

### Collett 3500-gallon

Built between April 1929 and March 1930 under Lot A118, Nos. 2242-68 were the most modern tenders known to have worked with the class. They closely followed the style of the Collett 4000-gallon tender but could be identified by the shallower body. The other distinguishing feature of this design was the continuous fender wrapping around the rear of the tender, as opposed to it tapering to the end of the tank side.

**Above: The neat *modern* lines of the Collett 3500-gallon tender with its wraparound fender can be appreciated in this view of No. 2916 *Saint Benedict* at Tyseley on 18 April 1936. The Engine Record Cards are far from conclusive when it comes to engine and tender pairings, but the earliest seen photograph showing a Saint with a Collett 3500g is March 1931. *Stephenson Locomotive Society.***

**Above: Collett 3500-gallon tender No. 2258 (Lot A118) adorned with the 'shirtbutton' monogram, is seen paired with No. 2980 *Cœur de Lion* alongside the coaling stage at Hereford shed. The date is 6 August 1939, less than a month before the outbreak of WW2.**

### Collett 4000-gallon

Although it has been annotated in the Engine Record Cards that No. 2931 *Arlington Court* ran with one of the largest of the Collett designed tenders (No. 2417) between September 1946 and September 1947, the research for this work has uncovered no photographic evidence to support this pairing.

**Dimensional summary of Saint class tenders :**

| Tender Type | Wheelbase | Wheel Diameter | **Weight** (tons) Full [Empty] | **Coal** (tons) |
|---|---|---|---|---|
| Dean 4000-gallon | 7' 6" + 7' 6" | 4' 0" later 4' 1.5" | 43.15 [19.25] | 7 |
| Churchward 3500-gallon | 7' 6" + 7' 6" | 4' 0" / 4' 1.5" | 40 [18.25] | 7 |
| Churchward 8-wheeled 3500-gallon | 5' 9" + 5' 6" + 5' 9" | 3' 2" | 45.75 [23.65] | 7 |
| Churchward 'Intermediate' 3500-gallon | 7' 6" + 7' 6" | 4' 1.5" | unknown | 7 |
| Collett 3500-gallon | 7' 6" + 7' 6" | 4' 1.5" | 43.15 [22.25] | 5.5 |

### Churchward / Didcot 3500-gallon

In order to complete the story of tenders allocated to the Saints, mention must be made of the hybrid vehicle created by the Great Western Society to run with No. 2999 *Lady of Legend*. To augment the early condition appearance of the locomotive, a new flush-riveted tank was constructed fitted with short fenders. This was mated with a reconditioned original chassis that the GWS had in stock (No. 1719), slightly out of period for the afficianardos as it features the wide-hornplates associated with later Collett era vehicles. Construction took place in the workshops at Didcot and the tender was completed in 2018. The outshopped tender was finished in the full **Great Western** lined green livery adorned with the company garter crest.

**Above: The new-build hybrid 3500-gallon tender is seen behind No. 2999 *Lady of Legend* at Didcot on 31 July 2021. *Hugh Llewelyn.***

# SAINTS AT WORK

As previously stated, the Saints were introduced for express passenger duties and were afforded that status until relatively late in their working lives. Under the GWR system they were designated as power classification C (tractive effort of 25,001–33,000 lbf) with Red route restriction (axle load up to 20 tons). This translated into power class 4P under British Railways.

Initial allocations reflected their intended use on fast heavy diagrams, with the class divided between London, Bristol, Plymouth, Newton Abbot, Cardiff and Goodwick. They soon established a reputation for good steaming and free running, with the large driving wheels permitting an impressive turn of speed. Fast and semi-fast top-link services became their 'bread and butter', including the non-stop 'Cornish Riviera Limited' between Paddington and Plymouth from 1904. Their haulage capabilities were regularly tested on the heavy Fishguard boat trains and from 1910 a demanding two-hour diagram between Paddington and Birmingham Snow Hill. On 9 July 1923 No. 2915 *Saint Bartholomew* earned the accolade of hauling the accelerated 2.30pm express from Cheltenham to Paddington, the famous 'Cheltenham Flyer'.

With an increase in the number of the four-cylinder Stars and the introduction of Collett's Castle Class from 1923, the Saints began to be assigned to more secondary passenger and parcels duties. As a result they were allocated more widely across the network, to sheds including: Banbury, Chester, Gloucester, Hereford, Leamington, Newport, Reading, Shrewsbury, Swindon, Taunton, Tyseley, Westbury and Weymouth.

Motive power shortages during the Second World War saw a brief resurgence in the class being used on top-link express trains. But the widespread introduction of the Hawksworth Modified Hall and County Class 4-6-0s, eventually relegated the remaining Saints that had passed into British Railways ownership, into the ignominy of operating local stopping passenger services and the occasional freight duties. As such the class finished their days working predominately in the Midlands and South Wales.

**Above: A charming rural scene is captured at Ashley Hill on 7 September 1929, as No. 2922 *Saint Gabriel* draws attention as it passes with a Bristol-Swansea express. The Landore allocated engine is paired with an 'Intermediate' Lot A112 tender** ***W H Stone.***

Top: No. 2933 *Bibury Court* appears to have overshot the platform-end signals, with an Up express at Bristol Temple Meads in this undated view.

The lead vehicle of the train is Dean 40' Passenger Brake Van No. 1028, built in February 1891 as part of Lot 550 to Diagram K4. It was later re-categorised Diagram K30 following removal of its guard's lookouts, and was withdrawn in April 1939. It appears in good condition, retaining lined panelling. In the background, the Iron Mink also looks in good condition with 16" GW lettering that was introduced in 1922.

***Milepost Collection.***

Middle: In May 1933, No. 2973 *Robins Bolitho* is captured replenishing its tanks on the Goring Water Troughs, whilst working a milk train on the Down Main line. Entering service in March 1905, the locomotive was to be withdrawn two months after this scene was captured.

By this date there were 126 members of the Hall Class 4-6-0s in service, and their growing numbers were encroaching on slower passenger services. Accordingly, Saints were beginning to operate secondary traffic such as stopping passenger, parcels and milk trains.

***Dr Ian C Allen.***

Bottom: No. 2924 *Saint Helena* brings a Down express over Red Cow Level Crossing and into Platform 1 at Exeter St Davids station.

The train seems to be an inter-company working with the first vehicle a 57' flat ended Collett Brake Composite (introduced in 1923). These versatile coaches were often diagrammed to work into 'foreign' territory, and as such could be marshalled into mixed return formations. The second vehicle appears to be an ex-LNWR elliptical-roofed Brake Third, and the next an ex-MR or early LMS designed coach.

***Arthur Mace (MC 20469).***

Above: No. 2946 *Langford Court* at Exeter St Davids with a Down express. The first vehicle is one of four Diagram K17 54' Clerestory Postal Brake vans (Nos. 1125-8), built in December 1903 for ordinary mail traffic and withdrawn 1951-4. *Milepost Collection.*

Below: Capturing pre-Nationalisation trains at night was a challenging task due to contemporary film speed and camera shutter limitations, so this nocturnal photograph of No. 2914 *Saint Augustine* at Birmingham Snow Hill is quite a rarity. *Milepost Collection.*

The photographs over these two pages all show locomotives on shed during the mid-1930s, but with various detail differences. No. 2977 *Robertson* rests by Old Oak Common coaling stage in July 1934. It retains the as-built straight running plate and exposed valve spindle covers, but has acquired a short safety valve cover and ATC shoe. *Milepost Collection.*

Above: Still ostensibly in as-built condition, No. 2931 *Arlington Court* is seen at Wolverhampton Stafford Road shed on 29 August 1936. The only discernible external modification appears to be the repositioning of the top lamp bracket onto the smokebox door and the removal of the ineffective bogie brakes. As can be seen on these later batch locomotives, the access to the valve spindles was via a hinged cover which incorporated a small straight step. *W A Camwell (courtesy Stephenson Locomotive Society).*

Below: During a Heavy General overhaul in March 1936, No. 2979 *Quentin Dunward* received a modified front end which included extended frames, new cylinders and outside steam pipes. This substantial work was undertaken on 39 Saints between 1930 and 1948. In this form it stands at Cardiff Canton in 4 October 1936. Note the box style valve spindle covers on the front fall-plate, retained tall safety valve bonnet and the 'shirtbutton' tender monogram. *W A Camwell (courtesy Stephenson Locomotive Society).*

This colourised black and white original photograph shows No. 2989 *Talisman* awiting the 'right away' from Leamington Spa, *circa* 1935. *David Williams*.

No. 2949 *Stanford Court* has just been replenished on the coaling road at Shrewsbury shed. This colourised view is thought to be *circa* 1937 when the locomotive was paired with a Collett 3500g tender. *David Williams*.

Above: The crew and a curious young observer await departure from Leamington Spa with a Birmingham express behind No. 2915 *Saint Bartholomew.* This locomotive was one of 37 members of the class to never receive outside steam pipes. *Arthur Mace.*

Below: No. 2928 *Saint Sebastian* rolls into Salisbury with a train bound for Portsmouth on 5 August 1947. Another example of the class to retain its original cylinders until withdrawal from Westbury shed in August 1948. *Peter Pescod (PP 353).*

Other than the addition of a smokebox door numberplate and 86C (Cardiff Canton) shedcode, not much of the 'old order' appears to have changed in this post-Nationalisation view of No. 2940 *Dorney Court* at Bridgend on 5 May 1951. The Collett 3500-gallon tender retains the post-war G (Crest) W emblem, but close inspection suggests the name and number plate are red, suggesting the loco at least is painted black. The leading coach is a Churchward 70' Toplight Gangwayed Third (Dia. C38) built 1920/21. *R C Riley (RCR 3095).*

**Above: The much photographed west end of Newton Abbot station, with No. 2927 *Saint Patrick* departing with a train for Paignton on 2 June 1951. The inclusion of ex-LMS coaching stock suggests a Summer Saturday inter-regional holiday service. *Peter W Gray*.**

**Below: An imposing viewpoint of No. 2951 *Tawstock Court* is captured on 1 August 1951, as it stands awaiting departure from Neath. Note the speedometer adjacent to the rear driving-wheel, which some of the class received from 1937. *R C Riley (RCR 3334)*.**

Above: No. 2927 *Saint Patrick* approaches Port Talbot on 4 August 1951, with an eleven-coach football excursion train from Swansea to Stoke-on-Trent. The locomotive was working out its final year, being withdrawn in December 1951. *R C Riley (RCR 3355).*

On 13 May 1952, No. 2937 *Clevedon Court* has arrived at Cardiff General with a stopping train from Hereford. The tender is a Collett 3500-gallon type. *R C Riley (RCR 3530).*

This Page: On 17 May 1952, photographer R C 'Dick' Riley procured himself a footplate ride between Cardiff and Hereford on No. 2938 *Corsham Court*. His record of the trip provides a unique viewpoint on operating a Saint, with a portrait of the crew prior to departure from Cardiff General (top) and 'on the road' north of Abergavenny, with Bryn Awr in the distance (bottom). R C Riley (RCR 3655/ 3656).

On Saturday 26 July 1952, No. 2906 *Lady of Lynn* stands prepared for duty alongside its home shed of Cardiff Canton (86C). By this date the locomotive had become the oldest surviving member of the class, and as such had attained celebrity 'pet' status amongst the depot staff. It retained some of the original Saint appearance, with the straight rear running plate and horizontal reversing lever. Despite the well cared for appearance, it was to only remain in traffic for less than a month, being withdrawn on 16 August. *R C Riley (RCR 3770).*

Above: Having now made its way to Cardiff North Yard, *Lady of Lynn* backs down to collect a train of loco coal. *R C Riley (RCR 3773).*

Below: Photographs of Saints on freight duties are rare, even in their later years. This next photograph in Dick Riley's sequence shows *Lady of Lynn* traversing the connecting line from Leckwith South Junction (on the route to Radyr) with Leckwith North Junction on the South Wales mainline. The load comprises fourteen loaded wagons, plus a Diagram AA20 20-ton Toad. The wagons mainly consist of an eclectic mixture of aged vehicles typical of the early 1950s, and can be identified in order from the tender as:- *(1 & 2)* ex-private owner 7-plank 10-ton, *(3 & 4)* GWR 20-ton Loco Coal, *(5)* 20-ton GWR Pole, *(6 & 7)* ex-PO 7-plank 10-ton, *(8 & 9)* ex-PO 8-plank 8-ton, *(10)* ex-PO 6-plank, *(11)* BR standard 16-ton mineral, *(12 & 13)* ex-PO 8-plank, *(14)* ex-PO 7-plank. *R C Riley (RCR 3774).*

LADY OF LYNN
2906

The 'Grand Old Lady' still ably earning her keep as the last survivor of the second production series of the class. No. 2906 *Lady of Lynn* powers a motley collection of vintage mineral wagons past Leckwith North Junction to the west of Cardiff on 26 July 1952. The GWR emblazoned Churchward 3500g tender and retained straight rear running plate create a scene from a bygone era. *R C Riley (RCR 3777).*

The tranquil surroundings of Llanvihangel on the Welsh Marches line is momentarily disturbed by the passing of No. 2945 *Hillingdon Court* on 1 June 1953. The train is a Cardiff General to Hereford stopping service, which was latterly a regular turn for a Saint from Cardiff Canton or Hereford sheds. This was to be one of the last duties for No. 2945, as it was withdrawn days later. *R C Riley (RCR 3907).*

# SAINTS IN DECLINE

The draw-down and demise of the Saints can be attributed to several factors, not just mechanical, but also economic, geo-political and business related. The engineering fragility brought on by age and the non-standard diversity of the class was an inevitable outcome, as was the continuous advancement in locomotive design and capabilities. But changing traffic demands, economic pressure, the intervention of a second world wide conflict, and a politically motivated Nationalisation of Britain's railways all contributed to the decline and eventual extinction of Churchward's masterpiece.

For many years there was much lamenting amongst professional railwaymen and the enthusiast fraternity that a Saint did not warrant the priority for preservation that the class richly deserved. Fortunately, this oversight has now been rectified, as discussed in the final chapter of this work.

## Inter-War Years (1922-1939)

To the regret of many observers who thought the action premature, withdrawal of Saints on grounds of age or condition commenced in 1931 (and of 4-6-0 Stars soon after). The context for these changes is best judged within the replacement programme initiated soon after the Grouping which was of a scale comparable with that which had stemmed from the Dean-Churchward transition of around twenty years earlier. The key difference was that there was little of the revolutionary element that had so distinguished the work of Collett's predecessor.

Taking to account the 20th Century locomotives (plus a few from the late 19th Century), the evolutionary pattern of the tender locomotive fleet with capacity to handle mainline express or mixed traffic services is summarised:

| Year | 4-4-0 (all) | 43xx 2-6-0 | 29xx 4-6-0 | 49xx 4-6-0 | 68xx 4-6-0 | 78xx 4-6-0 | 40xx 4-6-0 | 4073 4-6-0 | 60xx 4-6-0 | Total 4-6-0 | Total (all) |
|---|---|---|---|---|---|---|---|---|---|---|---|
| **1922** | 392 | 373 | 77 | - | - | - | 73 | - | - | 150 | 915 |
| **1923** | 397 | 394 | 77 | - | - | - | 73 | 2 | - | 152 | 943 |
| **1924** | 393 | 404 | 76 | 1 | - | - | 73 | 11 | - | 161 | 958 |
| **1925** | 387 | 404 | 76 | 1 | - | - | 71 | 23 | - | 171 | 962 |
| **1926** | 383 | 404 | 76 | 1 | - | - | 69 | 35 | - | 181 | 968 |
| **1927** | 344 | 404 | 76 | 1 | - | - | 69 | 45 | 6 | 197 | 945 |
| **1928** | 315 | 404 | 76 | 6 | - | - | 69 | 45 | 20 | 216 | 935 |
| **1929** | 276 | 404 | 76 | 71 | - | - | 68 | 46 | 20 | 281 | 961 |
| **1930** | 250 | 403 | 76 | 82 | - | - | 68 | 46 | 30 | 302 | 955 |
| **1931** | 206 | 403 | 73 | 121 | - | - | 68 | 46 | 30 | 338 | 947 |
| **1932** | 194 | 423 | 69 | 121 | - | - | 66 | 56 | 30 | 342 | 959 |
| **1933** | 185 | 423 | 63 | 141 | - | - | 65 | 56 | 30 | 355 | 963 |
| **1934** | 162 | 423 | 59 | 141 | - | - | 60 | 66 | 30 | 356 | 941 |
| **1935** | 147 | 388 | 55 | 158 | - | - | 58 | 76 | 30 | 377 | 912 |
| **1936** | 126 | 349 | 55 | 166 | 20 | - | 58 | 91 | 30 | 420 | 895 |
| **1937** | 116 | 311 | 55 | 176 | 60 | - | 55 | 104 | 30 | 480 | 907 |
| **1938** | 105 | 290 | 55 | 186 | 60 | 12 | 52 | 116 | 30 | 511 | 906 |
| **1939** | 93 | 271 | 54 | 195 | 80 | 8 | 49 | 130 | 30 | 546 | 910 |

Several features deserve comment:

1) The reign of the large wheeled (6' 8½") 4-4-0s as top link express locomotives was brief and their overall careers had been short:

| Class | Introduced | Last in Service |
|---|---|---|
| **Armstrong** | 1894 | 1930 |
| **Badminton** | 1897 | 1931 |
| **Atbara** | 1900 | 1931 |
| **City** | 1903 | 1931 |
| **County** (4-4-0) | 1904 | 1933 |
| **Flower** | 1908 | 1931 |

2) The surviving 4-4-0s comprised classes Bulldog, Duke and Earl (32xx) whose 5' 8" driving wheels ensured continued use into BR days primarily over secondary routes.

3) The persistent trend toward locomotive enlargement was largely caused by increasing passenger train loads through rolling stock modernisation. Replacement of timber-bodied 50-60' coaches (Clerestory and Toplight types) with modern steel-bodied stock significantly increased the tare per seat factor. The growing presence of heavier mixed traffic category trains demanded smaller-wheeled 4-6-0s, as initiated by the Saint class's offspring (No. 2925/ 4900) which provided the Hall Class prototype in 1924.

4) Reports from the early 1930s showed that 2-6-0 Class 43xx fitted with the Standard No. 4 boiler was increasingly stretched on some passenger services. This situation heralded the combination of 5' 8" driving wheels and Standard No. 1 boiler in the Grange 4-6-0s as replacement for the Moguls. This development belatedly filled a need foreshadowed by Churchward in his motive power plan of 1901.

5) The comprehensive inter-war up-grade/ enlargement programme involved withdrawals, rebuild and new construction on a substantial scale. Although outside the brief of this issue, the exercise must have incurred considerable financial cost with new locomotives replacing earlier machinery on virtually a one-for one basis as supported by the near constant aggregate fleet totals which varied between 895 and 968 (average 940) over the 18-year period. Thus there appears to have been no effort to contain the aggregate cost by numerical fleet reduction through improved utilisation, a measure that proved a highly successful component of William Stanier's mighty re-stocking programme on the LMS.

**Above: Historically the first of the class, No. 2900 (formerly No. 100) *William Dean,* paints a sorry picture as it languishes in the Swindon Conyard on 10 September 1932. The engine had latterly seen service from Chester shed, from where it was withdrawn on 17 June 1932 and quickly transferred to Swindon for scrapping. It is recorded as having run 1,260,804 miles. What price now for the still attached name and number plates on the modern collectors' market?** ***W L Good (courtesy Stephenson Locomotive Society).***

### Withdrawals

The first Saint to be removed from traffic was No. 2925, albeit for rebuild in 1924 as the prototype for the Hall Class as discussed earlier. There were also twenty-two age or condition related withdrawals in the pre-World War 2 period, some due to cracked frames and in the case of No. 2900, likely due to its non-standard cylinders:

| Loco No. | Built | Wdn | Loco No. | Built | Wdn |
|---|---|---|---|---|---|
| **2900** | 1902 | 1932 | **2972** | 1905 | 1935 |
| **2901** | 1906 | 1933 | **2973** | 1905 | 1933 |
| **2904** | 1906 | 1932 | **2974** | 1905 | 1933 |
| **2907** | 1906 | 1933 | **2976** | 1905 | 1934 |
| **2909** | 1906 | 1931 | **2977** | 1905 | 1935 |
| **2910** | 1906 | 1931 | **2982** | 1905 | 1934 |
| **2911** | 1907 | 1935 | **2984** | 1905 | 1933 |
| **2917** | 1907 | 1934 | **2985** | 1905 | 1931 |
| **2918** | 1907 | 1935 | **2986** | 1905 | 1932 |
| **2919** | 1907 | 1932 | **2990** | 1905 | 1939 |
| **2923** | 1907 | 1934 | **2998** | 1903 | 1933 |

When taken out of service, all carried superheated full-cone boilers and retained the original front end with inside steam pipes (fitting of outside steam pipes commenced with Nos. 2905/ 12/ 18/ 35 in 1931). The average working career of this group was 28 years.

### Duties

The negative impact of the hammer blow issue, cessation of their construction in 1913, and continued introduction of 4-cylinder Stars followed by the 4-cylinder Castles saw fundamental change in the pattern of Saints' duties in the 1930s. They became more common on passenger services over less prestigious routes, plus deployment on semi-fast passenger, milk and parcels work. By 1936, No. 2937 *Clevedon Court* was reduced to standing as station pilot at Reading, but nonetheless succeeded in being mentioned in despatches on three occasion:

**– 7 April**: The 3.30pm Paddington-Penzance express was observed arriving at Plymouth about 5 minutes late behind No. 2937 working solus. This train was scheduled to stop at Taunton, Exeter and Newton Abbot (where the Torbay section was detached). The allowance for the Paddington-Exeter leg was 3 hours 5 minutes and assuming that the locomotive had been substituted for a failure at Reading, there must have been some exhilarating work to arrive at North Road only 5 minutes down.

**– Circa two weeks later**: In conjunction with the 1935 Centenary celebrations, a down duplicate service to the 'Cornish Riviera Express' departing at 10.35am had been instituted. The main train served various Cornish destinations while the duplicate catered for stations and routes previously reached by the CRE's slip portions. Thus the 10.35am train slipped the Weymouth section at Heywood Road Junction, then stopped at Taunton to detach coaches for Ilfracombe and Minehead, followed by Newton Abbot where a through coach for Kingsbridge was detached. A *Railway Magazine* correspondent travelling to Taunton on the 10.35am noted that the train slowed on approach to Reading and was diverted through the relief line platform before reaching the Berks & Hants route. While passing slowly through Reading, the CRE was noted stationary on the down main with its King in process of being replaced by station pilot No. 2937. The Castle hauling the duplicate recovered from the Reading delay and reached Taunton on time at 12.57pm (142.7 miles in 142 minutes) where the correspondent alighted; the duplicate departed at 1.0pm. The correspondent then waited to observe developments and a little after 1.05pm, No. 2937 stormed through Taunton going 'flat out'. Presumably the duplicate was held at Norton Fitzwarren for the CRE to overtake and resume the ordained order of services.

**– Later in the year**: No logs were preserved for these two journeys, but R E Charlewood (an experienced and meticulous timer) was aboard the 340-ton 'Bristolian'. He noted an unusually vigorous start from Paddington by No. 6015 *King Richard III* but speed had radically reduced by Maidenhead due to a hot box. Apparently it was possible to pass a message to a signalman warning of impending failure at Reading, which allowed about five minutes' notice to prepare the station pilot as substitute. The 'Bristolian' was then allowed 72 minutes between passing Reading and arrival at Temple Meads (82¼ miles, average speed 68.5 mph). Apparently No. 2937's qualities had become common knowledge as the driver meted out similar treatment to the substitute that he had given the King on leaving the capital. From a dead start, 78 mph was reached in 9 minutes and the flying average over 64½ miles from Pangbourne to MP 106 was 74½ mph. It was calculated that the net time Reading-Bristol equated 66½ minutes and a pass-to-stop average speed of 74 mph.

**The first official withdrawal for scrapping was No. 2985 *Peveril of the Peak*, which was condemed from Landore on 8 August 1931. Seen here at Nottingham Victoria *circa* 1921. *T G Hepburn.***

### World War 2 and its Aftermath (1940-1947)

The challenging operating conditions meant retention of all serviceable locomotives and withdrawals only resumed in 1944 as follows:

| Loco No. | Built | Wdn | Loco No. | Built | Wdn |
|---|---|---|---|---|---|
| 2914 | 1905 | 1946 | 2975 | 1905 | 1944 |
| 2921 | 1905 | 1945 | 2978 | 1905 | 1946 |
| 2922 | 1905 | 1944 | 2983 | 1905 | 1946 |
| 2971 | 1903 | 1946 | (Careers average – 41 years) | | |

Relaxed schedules during the conflict rendered the hammer blow penalty irrelevant, while the superior starting and hill climbing characteristics imparted by Stephenson Link Motion was a useful advantage. As was standard by this time, all carried superheated full-cone boilers while retaining the original inside steam pipe front end.

### Survival into British Railways' Days

Of the original 77 locomotives, one had been converted to the Hall prototype, twenty-two had been withdrawn and scrapped in the period 1931-9, and seven had been withdrawn by the GWR during the company's final five years of independent existence. This left 47 class members to be inherited by the new owner on 1 January 1948. Despite the withdrawals thus far, the class's heterogeneous outward appearance was still evident among these survivors.

### 'Ladies' Series (between Nos. 2902 and 2908)

The ten Ladies were built with right-angled/ vertical front fall and valences, screw reverse, straight running plate from smokebox to the cab rear, and prominent cab footsteps. Apart from pre-World War 2 changes in boiler type (half/ full cone) and top feed installation, Nos 2901/ 4/ 7/ 9/ 10 remained unchanged until withdrawal between April 1932 and July 1933. The remaining five were all fitted with outside steam pipes, concurrent with fitting of curved front fall and valences similar to the Saint and Court series as built. The original straight running plate and footsteps were retained at the rear end plus, rather curiously, the lever (pole) reverse thus rendering a hybrid appearance. Photographic evidence has been traced to confirm this condition on all but No. 2908 which most probably conformed with the others.

| No. | Outside Steam Pipe Fitted | Withdrawn |
|---|---|---|
| 2902 | Dec-43 | Aug-49 |
| 2903 | Nov-30 | Nov-49 |
| 2905 | Mar-31 | Apr-48 |
| 2906 | Oct-35 | Aug-52 |
| 2908 | Dec-35 | Dec-50 |

### 'Saints' Series (between Nos. 2912 and 2930)

As the first of the 29xx series to be built with curved front and rear falls/ running plates, there had been less visual variety beyond boiler/ top feed changes. Eleven were still in service as at 31 December 1947 of which four retained original inside steam pipes.

| No. | Outside Steam Pipe Fitted | Withdrawn |
|---|---|---|
| 2912 | Apr-31 | Feb-51 |
| 2913 | - | May-48 |
| 2915 | - | Oct-50 |
| 2916 | Feb-43 | Jul-48 |
| 2920 | Dec-32 | Oct-53 |
| 2924 | Sep-33 | Mar-50 |
| 2926 | Jun-34 | Sep-51 |
| 2927 | Jul-45 | Dec-51 |
| 2928 | - | Aug-48 |
| 2929 | - | Dec-49 |
| 2930 | Sep-33 | Oct-49 |

### 'Court' Series (Nos. 2931-55)

All 25 of this series were built in the same style as the Saints and apart from five carrying half-cone boilers intermittently, throughout their careers they were the most homogeneous group appearance-wise. With two exceptions, all acquired outside steam pipes: No. 2947 as noted below and No. 2935 which retained its unique rotary cam poppet valves.

| No. | Outside Steam Pipe Fitted | Withdrawn |
|---|---|---|
| 2931 | Dec-37 | Feb-51 |
| 2932 | Dec-37 | Jun-51 |
| 2933 | Mar-42 | Jan-53 |
| 2934 | Nov-38 | Jun-52 |
| 2935 | May-31 (Rotary Valves) | Dec-48 |
| 2936 | Mar-32 | Apr-51 |
| 2937 | Jun-48 | Jun-53 |
| 2938 | Feb-35 | Aug-52 |
| 2939 | Nov-36 | Dec-50 |
| 2940 | Oct-43 | Jan-52 |
| 2941 | Apr-32 | Dec-49 |
| 2942 | Apr-33 | Dec-49 |
| 2943 | Aug-33 | Jan-51 |
| 2944 | Feb-44 | Nov-51 |
| 2945 | Jan-38 | Jun-53 |
| Table Continued Overleaf | | |

| No. | Outside Steam Pipe Fitted | Withdrawn |
|---|---|---|
| **2946** | Jun-33 | Nov-49 |
| **2947** | - | Apr-51 |
| **2948** | Feb-32 | Nov-51 |
| **2949** | Dec-35 | Jan-52 |
| **2950** | Feb-34 | Sep-52 |
| **2951** | Aug-33 | Jun-52 |
| **2952** | May-32 | Sep-51 |
| **2953** | Apr-32 | Feb-52 |
| **2954** | Mar-33 | Jul-52 |
| **2955** | Jan-34 | Mar-50 |

**Ex-Atlantics (Nos. 2980-2, 2987-9)**

All of the remaining six locomotives had been built as 4-4-2s.

| No. | Built | Converted to 4-6-0 | Outside Steam Pipe | Wdn |
|---|---|---|---|---|
| **2979** | Apr-05 | Aug-12 | Mar-36 | Jan-51 |
| **2980** | May-05 | Jan-13 | - | May-48 |
| **2981** | Jun-05 | Jul-12 | - | Mar-51 |
| **2987** | Aug-05 | Jun-12 | Mar-42 | Oct-49 |
| **2988** | Aug-05 | May-12 | - | May-48 |
| **2989** | Sep-05 | Oct-12 | - | Sep-48 |

In this condition they closely resembled the Court series but for retention of lever reverse and the horizontal rod set outside the line of the splashers.

## Summary

The period in public ownership saw a pattern of employment similar to that of the 1930s, and their demise was mainly caused by entry into service of the Hawksworth Modified Halls Nos. 6971-99/ 7900-29. Recorded career mileages varied between 1,156,000 and 2,076,000; this wide spread is understandable given the prolonged withdrawal period which commenced in August 1931 with No. 2985 *Peveril of the Peak* and concluded in October 1953 with No. 2920 *Saint David.*

**Below: The end of the road for a faithful old servant. No. 2981 *Ivanhoe* stands in the Swindon Conyard alongside 'C' Shop on 17 June 1951, awaiting its turn with the cutters torch. It had been withdrawn from Banbury in March. *Milepost Collection.***

Typical of the neglected condition the class endured during their final few years in traffic, September 1907 built No. 2926 *Saint Nicholas* stands in the shed yard at Chester (84K) on 21 July 1951. The local resident valiantly attempts to display its once proud British Railways lined black livery beneath the layers of grime, and just discernable is the GWR-style CHR stencil on the front running plate. *R C Riley (RCR 3228).*

Above: Devoid of its smokebox numberplate and looking in extremely unkempt external condition, No. 2954 *Tockenham Court* awaits the call at Carmarthen shed on 9 September 1951. Withdrawal for this penultimate-built Saint came in July 1952. *R C Riley (RCR 3445).*

Below: Whilst at first appearing to be withdrawn, No. 2945 *Hillingdon Court* is in fact receiving some running repairs to its pistons in the Swindon Stock Shed on 15 June 1952. It was to remain operating from Cardiff Canton for another 12-months. *R C Riley (RCR 3689).*

# SAINT DAVID
## THE ONE THAT GOT AWAY

Resplendent in BR lined black livery, No. 2920 *Saint David* rests between duties at its home shed Hereford (85C), on 17 May 1952. This was the start of a period of 'celebrity' for the locomotive working enthusiast specials. *R C Riley (RCR 3659).*

Above: By the time of this photograph of a work-stained *Saint David* at Worcester Shrub Hill on 24 July 1951, only 18 of the class were still active, operating from Bath Road, Canton, Chester, Gloucester, Hereford, Leamington and Swindon. *R C Riley (RCR 3261).*

Below: On Sunday 15 June 1952 the Stephenson Locomotive Society (SLS) ran their 'Saint Special' from Birmingham Snow Hill to Swindon and return. Here, at around midday, No. 2920 prepares to run around the train at Gloucester. *R C Riley (RCR 3668).*

Shortly after the previous view, No. 2920 is caught taking a spin on the Gloucester shed turntable, in preparation to continue the journey to Swindon. *R C Riley (RCR 3671).*

After a visit to Swindon Works, the SLS 'Saint Special' departs past Rodbourne Lane Signal Box. The return journey was via Banbury and Leamington Spa, arriving back into Birmingham Snow Hill at 9.35pm. *R C Riley (RCR 3692).*

Top: By the start of 1953 only four Saints remained in revenue service, Nos. 2920 *Saint David* (Hereford), 2933 *Bibury Court* (Leamington), 2937 *Clevedon Court* (Hereford) and 2945 *Hillingdon Court* (Cardiff Canton).

*Saint David* continued to be rostered for special workings, although the paintwork lustre of the previous year was starting to wane. It is captured here in Cardiff Canton shed yard on 1 June 1953 coaled and ready for duty, with her footplate crew proudly posing with their mount.

*R C Riley (RCR 3891).*

Middle: Later the same day, the locomotive is seen in Cardiff Canton carriage sidings collecting the stock for what is likely an afternoon working to Hereford.

It made one final journey to Paddington on 27 August 1953 to work an excursion to Swindon, bringing to a close the Saints' long association with the capital. It is also recorded that the final few weeks in traffic were spent working Hereford-Worcester stopping trains.

By the time of its removal from service on 5 October 1953 after 46-years, No. 2920 had attained 2,076,299 miles run. This remarkable total was the second highest recorded mileage of any Great Western Railway locomotive, eclipsed only by No. 4037 *The South Wales Borderers*.

*R C Riley (RCR 3899).*

Above: A forlorn looking No. 2920 *Saint David* stands on the reception sidings at Swindon Works having been withdrawn with severely cracked frames. The date is 10 October 1953 and by mid-December it had been reduced to a pile of scrap in 'C' Shop. *Brian Wheeler.*

# PROGENY OF THE SAINTS

The Saints' pioneering role in development of the modern 2-cylinder 4-6-0 has been recounted many times. Indeed, W A Tuplin humorously describes the *Six-Foot Saints* (Hall), *Five-Foot-Eight Saints* (Grange), *Light-Weight Saints* (Manor) and the *High-Pressure Saints* (Hawksworth County). The class's acknowledged success is also considered to have fathered the LMS Black 5, LNER B1 and others; but less well-known are the chequered experiences that attended the first applications of Churchward's design philosophy beyond the GWR.

The story is complex and space permits only a summary of the details. In the early years of the 20th Century, Swindon recruited the services of E A Watson, a native of County Monaghan, Ireland who had previously worked with the Pennsylvania Railroad at Altoona and also the American Locomotive Company, Schenectady. This appointment was unusual as Swindon exported rather than imported senior personnel but it was characteristic of Churchward that with his interest in prevailing US practice, he sought to learn more through this means.

Watson worked as a subordinate of the Assistant Works Manager responsible for carriage and wagons matters and he might have influenced some designs that were reminiscent of US practice (e.g. Dreadnought coaches, Mink F bogie freight vans, and the gondola-like 40-ton locomotive coal wagons). Little is known of his background but he was unpopular with the rank-and-file for an abrasive manner and dictatorial management style. Also, his ambitions lay more in locomotive design and few tears were shed when in 1911 he left to become Works Manager under Richard Maunsell, Locomotive Superintendent of the Great Southern & Western Railway, Ireland.

Watson arrived at Inchicore thoroughly imbued with the Churchward design philosophy and Maunsell soon tired of his persistent lectures on Swindon's superiority. In 1913 Maunsell left to assume a similar position with the South Eastern & Chatham Railway and was succeeded by Watson who was as unpopular as his predecessor had been liked and respected. His dogmatic expression of political views contrary to prevailing nationalistic sentiment added to his unpopularity,

His large engine philosophy was evident in his two designs for the GS&WR, a solitary 4-8-0T heavy shunter in 1915 and the following year, 4-cylinder 4-6-0 Class 400 in which he sought to emulate the GWR Stars. Divided drive was employed but outside Walschaerts valve gear actuated the inside valves by means of rocking shafts. In full gear, the valve travel was 5¼" with a 1¼" lap, compared with the GWR standard of 1.63". Two years earlier, Maunsell had consulted E E Joynt, his erstwhile colleague at Inchicore who had followed Watson as Works Manager, about valve dimensions. Joynt, for all his other attributes, was of the old school and firm in the view that lap should be no more than 7/8". Unsurprisingly, Watson was at loggerheads with Joynt over the matter and 1¼" lap evidently resulted from a grudging compromise. This 'debate' shows how traditionalists had difficulties in accepting Churchward's revolutionary concepts.

A parallel Belpaire boiler with Schmidt superheater was installed on No. 400 which proved satisfactory as apart from the difficult start from Cork, mileages and gradients were less challenging than those between Paddington and Plymouth. Nevertheless, the ten locomotives eventually built encountered numerous problems and their popular appellation as the 'Irish Stars' was a travesty. They differed in several key details design-wise and above all in performance. R N 'Bob' Clements, doyen of Irish enthusiasts, studied the class's faults at length. He concluded that their root cause lay in flimsy mainframes and in sub-standard cross-bracing at their weakest point (i.e. around the cylinder area), which exposed Watson's slim experience in locomotive design and construction.

Dimensional comparison:

| **Company** | **GWR** | **GS&WR** |
|---|---|---|
| **Loco No.** (as built) | **4001** | **400** |
| **Cylinders** | (x4) 14.25" x 26" | (x4) 14" x 26" |
| **Walschaerts Valve Gear** | Inside | Outside |
| **Driving Wheels** | 6' 8.5" | 6' 7" |
| **Boiler Pressure** (lb/ sq in) | 225 | 175 |
| **Heating Surfaces** (sq ft) | | |
| - tubes | 1988 | 1614 |
| - firebox | 154 | 158 |
| - superheater | - | 440 |
| **Grate** (sq ft) | 27 | 28 |
| **Tractive Effort** (lbs) | 25,090 | 19,190 |
| **Weights** (tons) | | |
| - loco only | 75.6 | 70.7 |
| - adhesive | 55.4 | 50.7 |
| - max. axle loading | 18.6 | 17.3 |

**Above: The prototype of Great Southern & Western Railway 4-cylinder Class 400 as introduced in 1916. Conceptually a development of the Star Class with divided drive and outside valve motion, this type fell far short of the GWR 4-6-0 in both reliability and performance. Of the ten examples of Class 400, seven were rebuilt in three different forms and three had been withdrawn by 1930.**

Class 400's history in brief:

| No. | Built | Later History | Wdn |
|---|---|---|---|
| 400 | 1916 | Remained original | 1929 |
| 401 | 1921 | Rebuilt 1930 with 2-cylinders & Beardmore-Caprotti valve gear | 1961 |
| 402 | 1921 | Rebuilt 1927 with 2-cylinders & Walschaerts valve gear § | 1961 |
| 403 | 1923 | Rebuilt 1934 with 2-cylinders & Walschaerts valve gear | 1957 |
| 404 | 1923 | Rebuilt 1935 with 2-cylinders & Walschaerts valve gear as No.409 | 1958 |
| 405 | 1923 | Rebuilt 1933 with 2-cylinders & Walschaerts valve gear | 1955 |
| 406 | 1923 | Rebuilt 1930 with 2-cylinders & Beardmore-Caprotti valve gear | 1957 |
| 407 | 1923 | Rebuilt 1938 with 2-cylinders & Walschaerts valve gear | 1955 |
| 408 | 1923 | Remained original | 1930 |
| 409 | 1923 | Remained original | 1930 |

§ Rebuilt to Maunsell/ Holcroft specification; Nos. 403-5/ 7-9 built by Armstrong Whitworth; the identities of Nos. 404 and 409 were swapped unofficially at depot level as the latter was regarded as more deserving of a rebuild.

In a country famous for motive power longevity (e.g. 99-year career lifespan of GS&WR 0-6-0 Class 101 (J15)), early withdrawal of Nos. 400/ 4(original)/ 8 was a serious indictment. In an attempt to reduce problems, Nos. 407-9 had been built saturated with boiler pressure increased to 225 lb/sq in, an option soon rectified. Watson moved to become General Manager of Beyer Peacock in 1921, set himself up in fine style in a fashionable area of Manchester, only to succumb to cancer shortly afterward. His time at Swindon seems to have been air-brushed from history.

Watson's successor at Inchicore was J R Bazin who had trained at Doncaster. In desperation over the Class 400 calamity, he approached Maunsell at Ashford for help who in turn passed the problem to a recruit from Swindon, one Harry Holcroft. In 1927, his rebuilding specification was applied to No. 402. The boiler was still judged adequate but modifications below the running plate were extensive resulting in a new, sturdy 2-cylinder chassis.

**Above: No. 402 at Cork shed following its conversion from 4-cylinder propulsion in 1927. Among the 2-cylinder Class 400 rebuilds this locomotive was unique in acquiring a flush running plate from smokebox to cab.** ***C P Friel Collection.***

This treatment was expensive for the ever-impecunious Great Southern Railways (formed through the amalgamations of 1924/ 5). The later Caprotti conversions are believed to have been financed in part or entirely by Beardmore as a sales promotion. Rebuilding of the remaining quartet was cheaper yielding a less effective version of the Holcroft engine. The variations between the survivors demanded differing driving styles but footplate crews unanimously regarded No. 402 as the best of the seven. While the 'Irish Star' sobriquet for the 4-cylinder Class 400 was inappropriate, No. 402 may be considered the 'Irish Saint'.

**Above: The Irish Saint: Great Southern Railways 2-cylinder 4-6-0 No. 402 at Cork on an unrecorded date. Converted from an unsuccessful 4-cylinder in 1927 to a design prepared by Harry Holcroft (ex-Swindon drawing office), No. 402 was widely regarded by footplate crews as the best of the seven surviving 2-cylinder members of Class 400.** ***C P Friel Collection.***

Dimensional comparison:

| **Company** | **GWR** | **GS&WR** |
|---|---|---|
| **Loco No.** | **2955** (as built in 1913) | **402** (as rebuilt in 1927) |
| **Cylinders** | 18.5" x 30" | 19.5" x 28" |
| **Valve Gear** | Stephenson Inside | Walschaerts Outside |
| **Driving Wheels** | 6' 8.5" | 6' 7" |
| **Boiler Pressure** (lb/ sq in) | 225 | 180 |
| **Heating Surfaces** (sq ft) | | |
| - tubes | 1687 | 1590 |
| - firebox | 155 | 158 |
| - superheater | 287 | 350 |
| **Grate** (sq ft) | 27 | 28 |
| **Tractive Effort** (lbs) | 24,390 | 20,620 |
| **Weights** (tons) | | |
| - loco only | 74 | 73.5 |
| - adhesive | *circa* 54.8 | 54.85 |
| - max. axle loading | 18.6 | 19.5 |

The saga highlighted the importance of long valve travel. Bazin had accepted the point and his mixed traffic 2-cylinder 4-6-0 Class 500 with 5‘ 8½” driving wheels of 1924 was a superb design. Only three were built because kits for 12 moguls manufactured by the Woolwich Arsenal and ordered immediately pre-amalgamation by the Midland Great Western Railway had become available. Together with 15 more kits acquired later, 26 useful mixed traffic locomotives of classes 372 and 393 were eventually introduced at well below market cost. They were based on SE&CR Class N as Maunsell, in spite of his earlier resistance to Watson’s diatribes, had accepted the logic of the Churchward concept earlier than had Bazin. Also, Maunsell used long travel valves in his development of London & South Western 4-6-0 classes N15 and S15 from 1925 onward plus improvement of earlier Urie-built engines. Recognition of the Churchward design philosophy was thus gaining momentum beyond the Great Western Railway.

Sources: *Locomotives of the GSR* by Clements & McMahon, Colourpoint Books 2008.

*Locomotives of the Great Southern & Western Railway* by Clements, McMahon & O’Rourke, self-published 2020 (now distributed by Transport Treasury Publishing).

**Above: No. 500 was intended for mixed traffic duties but was equally at home on the prestigious Dublin-Cork ‘Enterprise’ express. Here it was leaving Mallow on the 11.50 am Cork-Dublin service on 16 June 1954.** ***B K B Green.***

# PHOENIX FROM THE ASHES

It has been a long-held ambition of the Great Western Society to represent as broad a spectrum of historical artefacts in the collection as possible. The absence of a Saint within the locomotive inventory was seen as a glaring omission, and a proposal to build a replica was first put to the GWS Board in 1971.

The plan was to reverse the Saint to Hall conversion of 1924, and thus the £4000 purchase of No. 4942 *Maindy Hall* from Barry scrapyard took place in 1974. The project finally began in earnest in late 1995, by which time initial fears of a lack of engineering capability to undertake such a project had been alleviated. New 6' 8½" driving and 3' 2" bogie wheels were produced and new cylinder blocks cast. The modified main frames and boiler from the Hall provided the basis for the new Saint, but a connecting rod from No. 2906 *Lady of Lynn* and the whistles from No. 2910 *Lady of Shalott* were also utilised. Whilst it has been built as a straight frame 4-6-0, rear extension frames to allow 4-4-2 configuration have been constructed, trial fitted and placed into store for potential future use.

Following thirty years of storage and fifteen years of rebuilding work, No. 2999 *Lady of Legend* made its first moves in April 2019 and was formally launched at the Didcot Railway Centre. The £825,000 project has been funded entirely by donations and bequests.

**In all its immaculate glory, No. 2999 *Lady of Legend* stands at Bridgnorth during a visit to the Severn Valley Railway on 11 June 2022.**

Above: The donor locomotive that made it all possible, No. 4942 *Maindy Hall* languishes at Woodhams scrapyard in Barry c1967. It arrived at the yard in 1964 and became the 51st locomotive to leave following purchase by the Great Western Society in April 1974.

Below: No. 4942 *Maindy Hall* stands on shed at Didcot in November 1995, on the launch day of the Saint Restoration Project. The locomotive was to donate its frames, boiler, smokebox, bogie frame and much of the motion parts to form the basis of the new build.

This Page: A selection of views taken on 28 July 2013 of No. 2999 *Lady of Legend* whilst under construction in the workshops of the Great Western Society at Didcot. The boiler was only temporarily placed in the frames at this time and was not completed for steam testing until late 2017. In March 2019 the locomotive steamed for the first time and the 48-year dream to recreate a working Churchward Saint had been finally realised. *Andrew Malthouse.*